Right Hemisphere and Verbal Communication

Yves Joanette   Pierre Goulet   Didier Hannequin

Right Hemisphere and Verbal Communication

With the Collaboration of John Boeglin

Springer-Verlag
New York Berlin Heidelberg
London Paris Tokyo Hong Kong

Yves Joanette, Ph.D.
Laboratoire Théophile-Alajouanine
Centre de recherche du Centre hospitalier
Côte-des-Neiges
Montréal, Québec H3W 1W5
Canada

Pierre Goulet, M.Sc.
Laboratoire Théophile-Alajouanine
Centre de recherche du Centre hospitalier
Côte-des-Neiges
Montréal, Québec H3W 1W5
Canada

Didier Hannequin, M.D.
Clinique neurologique
Centre hospitalier et Univ
Rouen
France

Library of Congress Cataloging-in-Publication Data
Joanette, Yves.
[Contribution de l'hemisphere droit a la communication verbale. English]
Right hemisphere and verbal communication/ by Yves Joanette, Pierre Goulet, and Didier Hannequin; with the collaboration of John Boeglin.
p. cm.
Translation of: La contribution de l'hemisphere droit a la communication verbale.
Bibliography: p.
Includes index.
ISBN 0-387-97101-7 (U.S.)
1. Language. 2. Cerebral dominance. 3. Aphasia. 4. Dyslexia. 5. Neurolinguistics. I. Goulet, Pierre. II. Hannequin, Didier. III. Title.
[DNLM: 1. Brain—physiology. 2. Dominance, Cerebral. 3. Language. 4. Language Disorders. WL 340 J62c]
QP399.J6213 1989
612.8′252—dc20
DNLM/DLC 89-19694

Printed on acid-free paper.

Typeset by Publishers Service, Bozeman, Montana.
Printed and bound by Edwards Brothers, Ann Arbor, Michigan.
Printed in the United States of America.

9 8 7 6 5 4 3 2 1

ISBN 0-387-97101-7 Springer-Verlag New York Berlin Heidelberg
ISBN 3-540-97101-7 Springer-Verlag Berlin Heidelberg New York

Foreword

Recently, the authors of this book reminded me that in 1959 I delivered a "seminal address on language dysfunction associated with right-brain damage" at a meeting of the American Speech-Language-Hearing Association (ASHA). In the light of this address, and subsequent publications on the subject, the authors asked whether I would be willing to write a preface to their book, *Right Hemisphere and Verbal Communication*. Of course, I was not only willing but delighted, more so since I had the opportunity to meet with one of the authors, Dr. Yves Joanette, 11 years ago, at a time when he only began to be interested in this topic. I was assured that I was free to write what I considered to be relevant on the subject. The result is what follows.

This book might well be entitled *The Right Hemisphere Rediscovered as to Its Role in Verbal Communication*. The book does not argue that recent research in any way suggests that the left hemisphere has lost competence, or dominance, for language functioning; it does argue that the left hemisphere is not alone in executing its responsibility *for all aspects of verbal communication*. Of course, the usual and well-established exceptions are recognized for some left-handers and ambidextrous, as well as the rare right-handed person who have right rather than left dominance for language. The investigations reviewed here emphasize the roles of each hemisphere to achieve efficient verbal communication.

When, as a doctoral student, I studied the anatomy and neurology of the brain's language system, I was impressed with the often-made observation that "the two brain hemispheres are as alike as any two peas in a pod." This was considered a truism to be accepted but not challenged by any actual observation. It was intended to imply that the left and right hemispheres of the brain, and certainly of what we could see of the cortex, were presumably identical in appearance and in anatomical structure. We students were therefore to join in puzzled wonderment the why of presumed anatomic symmetry in the light of the recognized functional asymmetry. Of course, we now know that the hemispheres are not symmetrical as Geschwind and Levitsky reported in 1968, based on their post-mortem findings on 100 adult brains. In a later review of his and other studies, in a 1979 article, "Specializations of the Human Brain,"[1] Geschwind reported

---

[1]N. Geschwind (1979), "Specializations of the human brain." *Scientific American, 241*, 180–199.

differences in cellular structure organization and in densities in the tissue layers of the cortex. Thus, the so-called mystery of assumed asymmetry of structure no longer holds. We have looked at open pea pods and found that the peas were not only different but that no two peas were quite alike. What, then, are the functional contributions of the brain to language and to linguistic communication?

At the very least, we should now be able to appreciate that linguistic usage in situational communication requires awareness of speaker intention and listener flexibility about possible intention beyond the literal and selected dictionary meaning or meanings of words. These communicative functions include nuances in voice changes, variations from usual prosodic patterns (syllable stress and word emphasis, as well as accompanying gestures and postural changes). We now have evidence that these subtleties in communication—semantic subtleties for high level meanings—may be impaired when the speaker or the listener has incurred damage to the right hemisphere. In general terms, we may conclude that, in most instances an undamaged right hemisphere is required for full and efficient verbal communication. This is true for full encoding and decoding of verbal messages alike.

This book sets "many things to right," including the historic, but largely ignored, contributions of Dr. Marc Dax, Dr. Paul Broca, and the observation of Dr. Hughlings-Jackson and the role he attributed to the right-brain hemisphere for the "control" of automatic (subpropositional) speech. Chapter 1 reviews the observations of neurologists who believed that the right hemisphere was not absolved of linguistic responsibilities.

I believe that we are not at a stage when we should consider the need to redefine aphasic and other language disorders. We need again to ask ourselves whether aphasia is an appropriate term for all language disorders associated with brain damage other than those of dementia or whether we had better consider differentiating aphasic and other language impairments for efficient language processing that have a neuropathological basis. We will have to face—and not for the first time—the chicken-and-egg relationship of language and cognitive decrements as it relates to brain damage. Should the concept of superordinary language that was proposed by Dr. Macdonald Critchley at the Johns Hopkins University Conference, held in 1961, be evaluated in the light of the studies reported in this book? This concept requires the normally right hemisphere to interact with the left hemisphere for complete and efficient verbal processing. I believe that the time has arrived for researchers and clinicians—and for all aphasiologists for that matter—to assess linguistic and communicative capacities and, if possible, to develop inventories that are not now included in any depth in our present aphasic evaluation tests and procedures. *Right Hemisphere and Verbal Communication* both answers and questions the relevant issues of the possible role of the right hemisphere to verbal communication.

Stanford University, Stanford, California Jon Eisenson

Contents

Authors

Pierre Goulet, M.Sc. — Assistant de recherche, Laboratoire Théophile-Alajouanine, Centre de recherche du Centre hospitalier Côte-des-Neiges, 4565 Chemin de la Reine-Marie, Montréal Département de Psychologie, Université de Montréal, Montréal, Québec H3W 1W5 Canada

Didier Hannequin, M.D. — Praticien hospitalier, Clinique neurologique, Centre hospitalier et universitaire de Rouen, 76031 Rouen, France

Yves Joanette, Ph.D. — Scientist, Canadian Medical Research Council; Professeur agrégé de recherche, Ecole d'Orthophonie et d'Audiologie, Faculté de médecine, Université de Montréal; Membre actif, Laboratoire Théophile-Alajouanine, Centre de Recherche du Centre hospitalier Côte-des-Neiges, 4565 Chemin de la Reine-Marie, Montréal, Québec H3W 1W6 Canada

Collaborator

John Boeglin, Ph.D. — Assistant Professor, Department of Psychology, University of Saskatchewan, Saskatoon S7N 0W0 Canada

Introduction

Since the initial stirrings of Marc Dax in the first quarter of the nineteenth century, clinical and experimental observations have successively denied and then reimplicated a particular role for the right "nondominant" hemisphere in verbal communication. With the increasing importance of the question of the functional lateralization of the brain for language, studies in this area have become numerous. Thus it appeared to us, faced with a wealth of facts, that it was time to make the point as to the evidence gathered up until now that would argue for a specific and demonstrated role of the right hemisphere in verbal communication.

Consequently, the main objective of this book is to present a critical review of the issues and the facts relating to the contribution of the right hemisphere to verbal communication. Instead of proposing an ambitious conception or model, we look for evidence from many different approaches to converge. Each of the facts gathered is evaluated in terms of the degree of confidence one should have in it given conceptual or methodological contributions or limitations. It must be stressed that the "right hemisphere" that is discussed here is the "standard" right hemisphere of a typically educated right-handed occidental adult speaking and writing a language that shares the basic characteristics of such Indo-European languages as English, French, or German and whose right hemisphere is traditionally considered *nondominant* for language. Indeed, as has been well known for more than a century, the basic component of verbal communication referred to as *language* is dependent on neurobiological substrata that are mostly represented in the left hemisphere. Thus this monograph does not consider issues concerning the particular role of the right hemisphere in certain non–right-handers—left-handers and ambidextrals—or in the context of such exceptional conditions as those resulting in so-called crossed aphasia. These issues are not pertinent to the present topic since, in both cases, the left hemisphere is no longer considered the one that is mostly responsible for language.

As mentioned, the general question addressed here pertains to the degree to which the contribution of the left hemisphere to verbal communication is *relative*. Indeed, although the predominant role of the left hemisphere in language is a well-established fact, this does not necessarily imply that the left hemisphere is solely responsible for all the aspects of verbal communication. The amount of evidence suggesting that the right hemisphere demonstrates some competency in certain aspects of language and that it does contribute to verbal communication is more than enough to warrant a thorough review of the current status of the issue. This evidence, much of it recent, is aimed ultimately at understanding how the brain makes use of the contribution of its two hemispheres to achieve efficient verbal communication.

The choice of the expression "verbal communication" instead of the word "language" in the title of this book emphasizes the fact that the effective use of language is not restricted to mastering linguistic skills such as those that are defined by the rules of syntax, morphology, or phonology. Effective verbal communication also requires that one master the large number of rules governing the relationship between language and the different context in which it is used, both at the expressive and receptive levels. For example, to understand the intention of a speaker, one must be able to take into consideration both the literal meaning of the message and its intonation, or prosody, while also noting the context in which the message is spoken. In many instances we will see that the integrity of the right hemisphere is necessary for such contextual use of language. In restricting our review to *verbal* communication, we will not examine the nonverbal aspects of communication such as gestures, which are important components of interindividual exchanges but which have already been dealt with in more specific publications (see Bruyer, 1983; Feyereisen & de Lannoy, 1985; Nespoulous, Perron, & Lecours, 1986).

The objective of Chapter 1 is to provide the reader with the historical as well as conceptual backgrounds necessary to fully benefit from the critical review presented in subsequent chapters. As will be seen, the question of the possible contribution of the right hemisphere to verbal communication was totally neglected at a certain time. Indeed, for most of the earliest aphasiologists, the right hemisphere had nothing to do with the communicative abilities of right-handed adults. The pioneer studies that challenged this longstanding postulate are reviewed, as are those studies that somewhat systematically documented the effects of right-brain damage on the communicative abilities of right-handed individuals. Finally, some of the general genetic, environmental, and developmental determinants of this contribution are examined.

Apart from the evidence issued from the study of aphasia and its recovery, much of the evidence published concerning the role of the right hemisphere in verbal communication has come from the study of three distinct groups of subjects: right-brain–damaged subjects, commissurotomized subjects, and normal subjects. The purpose of Chapter 2 is to discuss the limitations associated with each of these groups as well as the limitations of the methods that have been used with them. This chapter thus lays the foundation for the subsequent chapters. It insists particularly on the fact that our understanding of the contribution of the right hemisphere can come only from a set of converging evidence issued from different, and complementary, sources of information.

Chapter 3 presents evidence for the contribution of the right hemisphere from a series of unique studies of recovering aphasic patients. The specific contributions of dichotic listening studies are reviewed and discussed. However, other studies making use of different and complementary approaches also are reviewed. It is argued that this converging evidence tends to suggest a certain role for the right hemisphere in the process of recovery after aphasia following a left-hemisphere lesion.

Chapter 4 allots a considerable amount of space to studies that have reported an involvement of the right hemisphere in lexical semantics. This aspect of the contribution of the right hemisphere to verbal communication is probably the one that has been most studied, in both normal and pathological populations. This chapter not only examines those facts that indicate a genuine competence of the right hemisphere for the semantic processing of words but also reviews those other facts pointing to the effective use of this competence as an example of the contribution of the right hemisphere to verbal communication.

The contribution of the right hemisphere to the semantic processing of words has led some researchers to speculate about its possible involvement in a particular aphasic condition referred to as *deep dyslexia*. Chapter 5 examines the available facts indicative of such a contribution, which derive from its demonstrated competence for the semantic processing of words.

Chapter 6 deals with the contribution of the right hemisphere to the processing of suprasegmental aspects of speech, generally referred to as prosody. Among other things, studies that examined the existence of acquired disturbances of prosody after right-hemisphere focal damage are critically reviewed. These studies, along with others that looked at normal subjects' performance, are reviewed separately, depending on whether they concern emotional or linguistic prosody. It is argued, among other things, that the distinction made between emotional and linguistic prosody does not coincide with the right and left-hemisphere dichotomy.

Chapter 7 examines how concepts developed within the field of pragmatics have allowed a better understanding of the verbal communication disorders that can be seen in right-brain-damaged patients. More specifically, the possible contribution of the right hemisphere is investigated by reference to both the production and the comprehension of the multiple components of connected speech. At the same time, the particular role of the right hemisphere concerning the integration of contextually based information in producing or understanding connected speech also is reviewed.

As already mentioned, this book does not try to reduce the complexity of the question posed and thus does not propose a unitary and simplified conception of the role of the right hemisphere in verbal communication. There is no easy and clear answer to this question. In fact, throughout this monograph, the reader should become aware of the problems that arise in attempting to identify and specify the right hemisphere's potential and/or effective contribution. Each chapter of this monograph illustrates the limits of this question, albeit for a specific aspect of the problem. However, it is our hope that the reader will share with us our enthusiasm in dealing with such a complex, difficult, but captivating question which lays down some aspects of the mutual relations between the mind and the brain.

Acknowledgments. Many thanks to Colette Cerny for her invaluable contribution to the management of this enterprise. This book would not have been possible without the support of the Conseil des recherches médicales du Canada (PG-28) and of the Fonds de la recherche en santé du Québec (Y. Joanette) as well as the support of Professor M. Samson, chief of the Clinique neurologique de Rouen, France, to which D. Hannequin belongs. A preliminary French version of this book, *La contribution de l'hémisphère droit à la communication verbale* by Didier Hannequin, Pierre Goulet, and Yves Joanette, was published by Masson (Paris). It also was the object of the annual *Rapport de Neurologie* presented jointly by Didier Hannequin and Pierre Goulet at the Congrès de Psychiatrie et de Neurologie de Langue Française held in Bordeaux, France, on June 15–19, 1987.

The nonpareil contribution of Dr. John Boeglin, our human English dictionary, is to be especially emphasized. It is only because of his participation in our collective effort that this English version was completed. Despite the suffering he, as well as his wife Marie-Josée and his children, experienced in revising this text, we hope that the topic of this book is no longer a nightmare for him and that he will even regain interest in it.

1
The Question

The prelude to the issues that are examined in this book coincides with the arrival of a detachment of troops from Napoleon's armies in a small village of southern France. Indeed, when a certain Marc Dax began practicing as a military surgeon in a large building overlooking the village of Sommières, sometime in 1800, a large number of soldiers, as well as civilians, came to him with clinical signs of cerebral lesions subsequent to head injuries, many of which were war-inflicted. Between 1800 and 1811 three patients in particular were to draw his attention to the respective roles of the two cerebral hemispheres in language function. The first patient, seen in 1800, was a captain in the cavalry who, after being stabbed in the head, experienced word memory problems without loss of object memory. This first case immediately drew Dax's attention to the strange phenomenon of word memory loss. The second patient was seen in 1809: this case of word memory loss subsequent to a tumor of the left hemisphere once again drew Dax's attention. Meanwhile, in 1806 a famous naturalist of the time, a certain Broussonnet, reportedly experienced word memory loss following an apoplectic seizure. Unfortunately, Dax was unable to obtain any information as to which side of Broussonnet's brain was affected. It was not until 1811, while reading Cuvier's funeral oration in honor of Broussonnet, that Dax found out that Broussonnet's autopsy had revealed the presence of a large ulcer in the left hemisphere of this illustrious patient. Almost immediately, Marc Dax was fascinated by the lateralization of the lesions, to the left in each of these cases. As time went by, Dax accumulated more observations of word memory loss, noting each time which side of the brain was affected. In 1836, at the Congrès Méridional de Montpellier, Marc Dax presented the results of his systematic observations to his colleagues from southern France. In his speech, entitled "Lésions de la moitié gauche de l'encéphale coïncidant avec l'oubli des signes de la pensée," Dax presented the following conclusion:

Non que toutes les maladies de l'hémisphère gauche doivent altérer la mémoire verbale, mais que, lorsque cette mémoire est altérée par une maladie du cerveau, il faut chercher la cause du désordre dans l'hémisphère gauche et l'y chercher encore si les deux hémisphères sont malades ensemble. (Dax, 1865, p. 261)[1]

This marked the first time that language behavior was associated with the left hemisphere and that the right hemisphere was implicitly discharged of any direct involvement in verbal communication. However, this paper was never officially published despite the efforts of Marc Dax and his son Gustave to provide copies to some colleagues (Caizergues, 1879).

It was not until May 1863 that Gustave Dax submitted a manuscript based on his father's presentation to the Académie de Médecine in Paris. A copy of this manuscript was later found by a grandchild of the dean of the Faculty of Medicine of Montpellier in office at the time, Professor F.-C. Caizergues. This copy was recognized by Gustave Dax as being one of the handwritten manuscripts he had helped his father distribute to his colleagues (Caizergues, 1879). At about the same time, a young surgeon was attracting the attention of the scientific world with his case studies of aphemia, or global aphasia, at the meetings of the Société d'Anthropologie de Paris. He presented several cases of language loss subsequent to lesions of the left hemisphere, but he refused to take a stand concerning the particular role of this hemisphere with respect to language control and, consequently, the absence of the right hemisphere's role. During the next two years, Paul Broca carried on with his observations, while the members of the Académie de Médecine disputed the originality of Marc Dax's manuscript. The debate surrounding this issue became increasingly intense. It reached a climax toward the end of the spring of 1865, when the members of the Académie de Médecine unanimously agreed on the authenticity of Marc Dax's paper, which had been presented in Montpellier some 31 years earlier. The French scientific world then turned its attention to Paul Broca. According to Ombredane (1951), Dax's manuscript was published only on May 25, 1865. Still according to Ombredane, Broca could no longer wait and had to reach a decision concerning the particular role of the left hemisphere with respect to the control of language behavior. He did so within 21 days and on June 15, 1865, at a meeting of the Société d'Anthropologie de Paris, declared:

De même que nous dirigeons les mouvements de l'écriture, du dessin, de la broderie, etc., avec l'hémisphère gauche, de même nous parlons avec l'hémisphère gauche. (Broca, 1865, in Hécaen & Dubois, 1969, p. 114)[2]

This time, the whole world heard the message. Whether it was due to Marc Dax's extraordinary sense of observation or to Paul Broca's dramatic statement, never again would the respective roles of the two cerebral hemispheres be con-

---

[1] "Not that all left hemisphere diseases necessarily impair verbal memory, but that, when this memory is impaired by brain disease, the cause of the disorder should be sought in the left hemisphere, and still sought there if the two hemispheres are both diseased" (authors' translation).

[2] "As much as we direct movements of writing, drawing, embroidery, etc. with the left hemisphere, so do we speak with the left hemisphere" (author's translation).

sidered as equal with respect to the main component of verbal communication—language.

This book is precisely about the possible role in verbal communication of the right hemisphere, the same hemisphere that was initially disregarded. More than a century has passed since the question of the possible role of the right hemisphere in verbal communication was raised. Although there are still no clear, straightforward answers to this question, it has been, and continues to be, rethought and reformulated with the steady accumulation of clinical and experimental findings. In this first chapter, we (1) describe some of the early studies concerning the question of a possible contribution of the right hemisphere to verbal communication, (2) comment on the studies of right-handed, right-brain–damaged populations that led to an initial formulation, albeit in general terms, of the question of right-hemisphere participation, and (3) examine the anatomical, genetic, environmental, and temporal determinants of this possible contribution. This chapter thus is intended to lead to a more precise formulation of the question that is the focus of this monograph.

The Pioneers

The teachings of Marc Dax and Paul Broca to the effect that the left hemisphere of the right-handed individual subtended the faculty of articulate language were clearly—and perhaps even too well—understood. Indeed, many of Broca's contemporaries scrupulously adhered to his teachings, including Mirallié (1896), who affirmed that "le cerveau droit n'a rien à voir avec les fonctions du langage" (p. 93).[3]

These teachings were passed on as is, or almost as is to the twentieth century, as attested to by Henschen (1926): "In speech, music and calculation, the left hemisphere plays a predominant or exclusive role" (p. 120). However, Broca's position was somewhat more moderate. Not only did Broca assign an important role to the right hemisphere of the left-handed individual, but he also recognized the possibility of a certain contribution of the right hemisphere of the right-handed individual to the recovery from aphasia in right-handers (see Chapter 3). But without a doubt, the writings of an English neurologist, John Hughlings-Jackson, were among the first in which a specific role was actually attributed to the right hemisphere, in coexistence with the predominant contribution of the left hemisphere.

Hughlings-Jackson first directed his attention to the various levels of language dissolution that occur in aphasia. Prompted by the writings of Baillarger (1865), Hughlings-Jackson was fascinated by the dissociation between impaired "volitional" behaviors and preserved "automatic" behaviors that occurred in most aphasics. This dissociation led him to formulate a hierarchical conception of the functional organization of the brain, with multiple levels of representation.

---

[3] "The right brain has nothing to do with language functions" (author's translation).

According to this theory, the role of the right hemisphere was considered to be necessary in order to organize the more volitional aspects of language – to "propositionalize," in Hughlings-Jackson's word – the right hemisphere was considered to be just as involved[4] in the more "automatic" aspects of language: "The right is the half of the brain for the automatic use of language, the left half for both the automatic *and* the voluntary use" (Hughlings-Jackson, 1915, p. 82, emphasis added).

Hughlings-Jackson's ideas were probably too divergent from the theories in vogue at the time. Among other things, his ideas differed in that they went beyond a simple Manichaean opposition between left and right, placing the respective roles of the two hemispheres within the context of a hierarchical organization. For this reason, Hughlings-Jackson's ideas went relatively unnoticed by his contemporaries. Had they been noticed earlier, it is highly probable that much of the early research efforts would have been more fruitful. But this was not the case. In fact, Hughlings-Jackson's theory fits in much better with contemporary theories as it insists, for example, on the importance of the dissociation between automatic and controlled processes.

From all of this emerged the classical view that the right hemisphere of the right-handed individual had little or no importance with respect to language function, a view that has been transmitted from one generation to the next. Occasionally, an illustrious personality has attempted to dispute this viewpoint. For example, at the turn of the century, Pierre Marie and Kattwinkel (1897) focused their attention on speech disorders in right- and left-hemiplegic patients, and came to the following conclusion:

> Nous bornerons à insister sur ce point, c'est que, contrairement à l'opinion généralement admise, les lésions de l'hémisphère droit jouent, dans la fonction du langage, un rôle dont on ne saurait méconnaître l'importance. (p. 518)[5]

The studies reported here were among the first to raise the question of a possible contribution of the right hemisphere to language, which was the only component of verbal communication recognized at the time, at least by neurologists. However, none of the studies, with the exception of those of John Hughlings-Jackson, provided a model of brain functioning in which the possible contribution of the right hemisphere could be accounted for. Be that as it may, it was not until much later – in fact, not until the middle of the twentieth century – that clinicians

---

[4]It is common to read in the literature that Hughlings-Jackson's concepts were based on an opposition between the left hemisphere's role with respect to the volitional or propositional aspects of language and the right hemisphere's role with respect to the automatic aspects of language. As shown here, such a conception is not that of Hughlings-Jackson. Indeed, he opposed the particular role of the left hemisphere to that of the two hemispheres considered together.

[5]"We will only insist on one point, that is, contrary to common belief, lesions of the right hemisphere do have a role in language function, a role whose importance should not be neglected" (author's translation).

began to examine this possible contribution, in particular the effects of acquired right-hemisphere damage on verbal communication abilities. Some of these early clinically oriented studies are briefly examined in the next section.

Early Studies

Historically, with the exception of a few sporadic reports describing the linguistic capacities of the right hemisphere in hemispherectomized patients (see Chapter 3), it was the observation of the linguistic capacities of right-handed, right-brain–damaged patients that led to the reconsideration of the question of a possible contribution of the right hemisphere to verbal communication. The findings of these early studies confirmed the presence of certain language disorders in right-handed, right-brain–damaged patients. In so doing, these studies supply the basis of many of the issues that are examined in this monograph.

Early studies of the linguistic capacities of right-handed, right-brain–damaged patients were not the only ones to inspire contemporary investigations. Subsequent studies of commissurotomized patients also played an important role in this respect. Nonetheless, the fact remains that the earliest studies to focus on the linguistic capacities of right-handed, right-brain–damaged patients addressed, though superficially, the issues that were to be subsequently reexamined. The language disorders that appear following a right-hemisphere lesion obviously cannot be qualified as being strictly aphasic in nature. It is difficult to analyze these disorders, mainly because of their discrete nature and the multiplicity of the factors involved. In dealing with his own difficulties following an infarct to the right hemisphere, the Norwegian anatomist Alf Brodal (1973) noted:

> There was a marked reduction in the powers of concentration which made mental tasks far more demanding than before. Reading novels did not cause great problems, but it was quite difficult and needed much concentration to follow the arguments, for example in a scientific paper. In part it seemed to be due to a reduced capacity to retain the sense of a sentence long enough to combine it with the meaning of the next sentence. (p. 685)

> Changes in writing and speaking like these may perhaps be classified as signs of dysgraphia and dysphasia . . . , but it appears that according to current views disturbances as those described are assumed to be caused by affections of the "dominant" hemisphere. (p. 686)

One can conclude from Brodal's observations that he was having difficulties with certain complex activities (e.g., relating different arguments in a scientific article) and that some of his errors were similar to those found in aphasia.

It should be noted, that in order to demonstrate the presence of a language disorder in a right-brain–damaged patient, one must use tasks that involve a level of functioning other than what is involved in the tasks that are typically employed in an aphasia battery. Generally speaking, the tasks should be more complex, though "complex" refers not just to the complexity of the stimuli used in a given task but also to the nature of processing required by that particular task. The fact of the matter is that adequate performance on a standard aphasia battery is not proof in itself of an adequate mastery of language in all of its aspects and certainly provides very little information concerning a particular individual's ability as an efficient communicator. What prompted researchers such as Eisenson (1959a, 1962) and Weinstein (1964) to study verbal communication abilities in right-handed, right-brain–damaged patients was the observation that although these patients were not aphasic, they obviously had communication disorders, or at least complained, as did Brodal (1973), of difficulties with respect to particular uses of language.

Moreover, simply identifying errors in verbal or written expression is not enough to prove that these errors are similar in nature to actual aphasic errors. Indeed, language constraints themselves can account, at least in part, for the particular surface form of an error. In these cases, when only general descriptive terms are used, the expression of problems at different levels may appear as similar. For example, normal language contains slips of the tongue, which, in other circumstances, could be considered as phonemic paraphasias (Fromkin, 1971). Although it is possible to observe errors in right-brain–damaged patients that are qualitatively similar to those observed in aphasics, quantitatively there are fewer (Joanette, Lecours, Lepage, & Lamoureux, 1983; Marcie, Hécaen, Dubois, & Angelerques, 1965). In any case, such a comparison has no explicative value at all. Studies of right-brain–damaged patients should not be limited to simply noting whether there are difficulties. Rather, their goal should be to understand the mechanisms that are at the origin of the difficulties through a better control of the experimental situations in which they are likely to appear.

The merit of some of the early investigators may well have been to go beyond the principle of left-hemisphere dominance to systematically examine the linguistic performance of right-brain–damaged patients. In the eyes of clinicians such as Eisenson (1959a, 1962) and Critchley (1962), it was obvious that these patients demonstrated difficulties that were unrelated to aphasic evaluation. Critchley, for example, described their difficulties in acquiring new linguistic information or in undertaking a creative literary task. To quantify these difficulties and go beyond a simple clinical impression, Critchley insisted on the necessity of using suitable tasks:

> Perhaps we still tend to overlook the possible role of the right half of the brain to the faculty of speech. Some aphasiologists are beginning to suspect that appropriate testing of a sufficiently searching character might well elicit defects within the sphere of language which are too subtle for ordinary routine techniques to bring to light. (Critchley, 1962, p. 211)

Archibald and Wepman came to a similar conclusion: "The errors of this group are, therefore, so subtle that they would not normally be noticed, except in a restricted test situation" (1968, p. 126).

However, not all the authors of this period agreed that if right-handed, right-brain–damaged patients did have language disorders, these were related to the more "complex" aspects of language. Two of the earliest studies illustrate this divergence of opinion. On the one hand, Eisenson (1962) supported Critchley's viewpoint that the right hemisphere contributes "superordinary" aspects of language, such as the ability to use abstract concepts. On the other hand, Marcie et al. (1965) attributed right-brain–damage impairments to the effect of disturbances of other cognitive functions on oral and written language, such as the presence of perserveration or inertia.

Eisenson (1962, 1973) compared the performance of 46 right-handed, right-brain–damaged patients and 46 normal controls on vocabulary tasks (e.g., word definition and recognition of the word corresponding to a definition) and sentence completion tasks. The percentage of correct definitions was similar in the two groups, but the number of incorrect words was significantly higher in the right-brain–damaged group. On the sentence-completion tasks, the missing word was either to be given spontaneously or to be chosen from among four alternatives. In both cases, the right-brain–damaged patients made more errors than the normals. Furthermore, these errors occurred for the most part in sentences that had to be completed with an abstract word. In light of these results, Eisenson concluded that the right hemisphere contributed to the more abstract aspects of language:

> As of now, our evidence is inconclusive. Tentatively, we might extend a speculation offered by Dr. Macdonald Critchley at a conference on Cerebral Dominance held at John Hopkins University on April 25, 1961. Dr. Critchley suggested that aphasia which usually occurred as a consequence of left-brain damage might be considered an impairment of ordinary language function. He speculated that the right cerebral hemisphere might have some higher than ordinary language potential. We believe that our findings may be interpreted to support Dr. Critchley's speculation. We suggest that the right cerebral hemisphere might be involved with super- or extra-ordinary language function, particularly as this function calls upon the need of the individual to deal with relatively abstract established language formulations, to which he must adjust. The modifications are not such as to be picked up in casual conversation, in which precision of language is not expected. (1962, pp. 52–53)

The Marcie et al. (1965) study came to an entirely different set of conclusions. These authors evaluated the performance of 28 right-brain–damaged right-handers (RBD), with various etiologies, on a series of written and oral language tasks typically used in the clinical evaluation of aphasia. The results of these patients were compared with those of a group of normal controls as well as those of a group of patients suffering from Parkinson's disease. The results can be summarized as follows:

1. With the exception of a slight tendency for dysprosodia (see Chapter 8), the RBD subjects did not show any particular disturbance of their conversational speech.
2. On word and "logatome" (nonword) repetition tasks, the performance of RBD subjects was somewhat similar to that of patients with so-called "milder forms of expressive aphasia." According to Marcie et al., the severity of these symptoms was between that of aphasic patients on the one hand and that of parkinsonian patients on the other hand.

On "lexical" (e.g., naming and verbal fluency) and grammatical tasks,[6] Marcie et al. reported that the performance of RBD subjects was more or less similar to the performance of normal subjects. The presence of perseveration was the only noticeable characteristic of the performance of RBD subjects on these tasks. According to Marcie et al., the main feature was the presence of a behavioral disorder described as an inertia, an inability to adapt to new instructions, expressed by a tendency to give perseverative responses.

4. On "written language" tasks (e.g., writing and reading), certain dysorthographic errors were observed in RBD subjects. According to the authors, graphic disturbances, such as letter and downstroke duplication, were the most noticeable.

In the next phase of their study, Marcie et al. (1965) proceeded to regroup the RBD subjects on the basis of their linguistic disorders and the results of a general evaluation of their cognitive functions. Despite methodological limitations (e.g., small number of subjects, various etiologies including pathologies with possible bilateral functional repercussions), five subgroups were identified. The first subgroup included oral and written perseveration along with dyscalculia and constructional apraxia. According to the authors, the lesion was usually located in the parietal lobe. The second subgroup was thought to be a reduced form of "motor" aphasia subsequent to a rolandic lesion. The third subgroup included repetition disorders, reading disorders, and dysorthographia. No particular lesion site was retained for this subgroup, although no occipital lesions were found according to the authors. The fourth subgroup included subjects demonstrating limited impairments with no common characteristics. According to the authors, the lesion usually affects the temporal lobe. Finally, the fifth subgroup included subjects with reading disorders subsequent to an occipital lesion. It should be noted, however, that these subgroups, as well as their anatomical correlations, have no statistical basis and reflect only the systematic application of a descriptive approach. Be that as it may, Marcie et al.'s conclusion focuses on the disturbance of "instrumental" language functions following a right-hemisphere lesion:

Cette première approche permet donc d'envisager la représentation sur l'hémisphère mineur de certaines fonctions instrumentales du langage, soit de manière très affaiblie

[6]The "grammatical" tasks included transformation of a sentence from singular to plural, transformation of a sentence from plural to singular, transformation of a sentence from masculine to feminine gender, transformation of a sentence from feminine to masculine gender, transformation of a sentence from present to past tense, transformation of a sentence from present to future tense, recognition of morphosyntactic errors, recognition of phonemic errors, recognition of semantic errors, forming semantic antonyms (e.g., good/bad), forming formal antonyms (e.g., cut/uncut), forming adverbs, verbal paradigms, sentence completion, generating sentences from words, and synonyms.

(réalisation motrice orale), soit dans l'arrangement spatial des signes écrits (écriture et lecture). (Marcie et al., 1965, p. 244)[7]

On the basis of the findings from these early studies, one can see that there was no unanimity concerning a clear formulation of the question of a possible contribution of the right hemisphere to language. For some, the question had to be formulated in terms of the contribution of certain "higher-level" aspects of language, whereas for others this possible contribution had to be restricted to the more instrumental aspects of language. This disagreement in the early literature probably reflects the incomparability of the clinical studies conducted at the time and the absence of any theoretical models to address this question. For example, the notion of "higher-level" or "superordinary" components of linguistic functioning, initially put forward by Jon Eisenson and MacDonald Critchley, does not correspond to any particular theoretical model; it was more the reflection of a clinical impression than one of objective and theoretically-based facts. Thus credit should be given to these early clinically-oriented studies for having drawn attention to the possible effects of a right-hemisphere lesion on language behavior, although none actually provided a satisfactory description. The question of the possible contribution of the right hemisphere to verbal communication had been formulated in general terms only, and much remained to be done.

Anatomical, Genetic, Environmental, and Temporal Determinants

The question of a possible contribution of the right hemisphere to verbal communication raises another set of issues: the anatomical, genetic, environmental, and even temporal determinants of this contribution. Although the identification of the exact role of these determinants has been the focus of attention in several studies of the left hemisphere's contribution to language, their role has received little, if any, consideration with respect to the eventual contribution of the right hemisphere. We examine this aspect of the question next, since it might eventually have a significant impact on many of the studies presented in subsequent chapters of this book, thereby leading to a more precise formulation of the original questions.

Let us consider, to begin with, the problem of functional intrahemispheric organization and the possible contribution of the right hemisphere to verbal communication. The following question arises: if the right hemisphere does indeed contribute to verbal communication, is this contribution aspecific or is it the

[7]"This first approach allows consideration of the representation of certain instrumental language functions in the minor hemisphere, whether in a diminished manner (oral motor realization) or in the spatial organization of written signs (writing and reading)" (author's translation).

result of a particular region, or of a particular neural network, which is topographically defined within the right hemisphere? Today, no one would go so far as to say that regardless of the region, the contribution of the left hemisphere is the same. However, nearly all of the studies of the possible contribution of the right hemisphere evade this aspect of the question. Considering once again the studies of right-brain–damaged patients, one way to operationalize the question would be to ask whether one should expect a right-hemisphere lesion to affect verbal communication, regardless of its locus, or whether one should expect a lesion of such-and-such a region of the right hemisphere to express itself by way of a specific disorder of verbal communication.

Generally speaking, much of the available evidence does not provide sufficient qualitative or quantitative information to deal with the issue of anatomoclinical correlations. Attempts to provide correlational data regarding certain disorders, such as those of prosody (Ross, 1981; Shapiro & Danley, 1985), are based on very small samples. In this regard, it should be noted that recent technological advances in medical imaging have led to a reconsideration of the classical relationships between lesions of the left hemisphere and aphasia type (Basso, Lecours, Moraschini, & Vanier, 1985; Poeck, de Bleser, & Keyserlingk, 1984). Even if anatomical evidence was available for the right hemisphere, the relevance of the resulting correlations would be limited. Aphasiological teachings are sufficiently adequate in showing that anatomoclinical correlations have no explicative value with respect to the mechanisms involved (e.g., agrammatism, deep dyslexia; Marshall, 1986). This lack of explicative value, however, does not mean that anatomical information should be disregarded altogether. Indeed, it is important to consider the topographical characteristics of the lesion when looking at the performance of right-brain–damaged subjects. For example, the conclusions of a study dealing exclusively with right-temporal–damaged patients should not be extrapolated to all right-brain–damaged patients (Tompkins & Mateer, 1985). By the same token, information concerning the actual locus of the lesion can possibly help explain some particular aspects of performance in terms of basic neurophysiology, without having to evoke more complex mechanisms. For example, the involvement of the left Heschl gyrus would be sufficient to explain dichotic listening performance in aphasia, without having to consider the possibility of a transfer of dominance to the right hemisphere (Niccum, Speakes, et al., 1986; see Chapter 6). As long as studies fail to provide specific information on the topography of the lesion, there is no clear way of formulating the question of a possible right-hemisphere contribution to verbal communication that could take into account the anatomical factor.

Still according to aphasiological teachings, there are other factors possibly related to a difference in the relative contribution of each hemisphere to linguistic functioning. Thus genetic factors such as sex (McGlone, 1980) and familial history of left-handedness or ambidexterity (Hécaen, Agostini, & Monzon-Montes, 1981) as well as environmental factors such as the language spoken (Paradis, 1983) and the exposure to schooling (Lecours et al., 1987) could also have an effect on the relative contribution of each of the two hemispheres to verbal communication. Ideally, then, these factors should be taken into consideration when formulating the question of a possible contribution of the right hemisphere to verbal communication. A study conducted by one of us (Y.J.) in collaboration with A. R. Lecours attempted to address this aspect of the question by describing the linguistic disorders, subsequent to a lesion of the right hemisphere, in a sample of right-handed individuals.

The goals of this study (Joanette, Lecours, et al., 1983) were threefold. The first goal was to describe the various language-related difficulties of right-brain–damaged subjects using a protocol that included a wide range of tasks. Generally speaking, these tasks were more complex than those usually administered in a standard evaluation of aphasia. The second goal was to identify subgroups within the total sample of right-brain–damaged patients based on the nature and the severity of their difficulties. The third goal was to examine any possible correlation between these subgroups and anatomical, genetic, or environmental factors.

On almost all of the 20 tasks used by Joanette, Lecours, Lepage, and Lamoureux (1983), the group of right-brain–damaged subjects ($N = 42$) performed significantly less than a group of control subjects consisting of 20 hospitalized patients free of any CNS lesion but comparable to the right-brain–damaged subjects in terms of age, sex, schooling, and knowledge of a second language. The tasks were (1) syllable, word, nonword, and sentence repetition, (2) word and sentence copying, (3) reading a text, (4) reading words such as *amidon* ("starch"), which can be read as two different words [ami ("friend")/don ("gift")], (5) spelling, (6) dictation, (7) naming objects and parts of objects, (8) verbal fluency according to semantic (e.g., animal, furniture) and orthographic criteria (e.g., words beginning with the letter *r* or *b*), (9) production of antonyms, (10) completing sentences with concrete or abstract words, (11) completing sentences with missing grammatical or lexical words, (12) oral definitions, (13) oral sentence construction using two or three words (e.g., intelligence, success, competition), (14) elaborating a story from a series of eight pictures, (15) reading comprehension, (16) verbal reasoning (e.g., "Paul is taller than Peter. Who is the smallest?"), (17) selecting a synonymous sentence from among four choices, (18) identifying a word (e.g., nature) presented as nonword (e.g., tanure) following the transposition of the two consonantal phonemes, (19) identifying a word inserted into a series of letters resulting in a nonword [e.g., "pont" ("bridge") in "brapontimo"], and (20) reading vertically presented words.

Apart from a dyslexia-dysgraphia of the spatial type, which was attributed to visuoperceptive deficits, the types of errors made by right-brain–damaged subjects on this battery were similar to those found in aphasia, at least at a surface-level description (e.g., phonemic paraphasias, word-finding difficulties, semantic paraphasias). To determine if there were any subgroups among the right-brain–damaged subjects, Joanette, Lecours, and colleagues (1983) used a descriptive method of multidimensional subject classification. This analysis led to the identification of subgroups. The first subgroup consisted of subjects who made very few errors. This subgroup included all of the normal subjects as well as nine right-brain–damaged subjects. The second subgroup consisted of 17 subjects with a moderate level of failure. Finally, the third subgroup consisted of the 16 least performing subjects.

According to this analysis, 33 of 42 right-brain–damaged subjects were found to be different from normal subjects. In the next phase of the analysis, information on each of ten parameters concerning anatomical, genetic, and environmental factors was collected for each of the brain-damaged subjects: pre- or retrorolandic lesion localization, cortical or subcortical involvement, lesion size, time since lesion onset, age, sex, familial history of left-handedness or stock handedness, level of education, and knowledge of a second language. A discriminant analysis of the data led to the identification of three parameters which accounted for the grouping of 70% of the subjects in a fashion similar to that obtained on the sole basis of the linguistic deficits: (1) the presence of cortical involvement (anatomical factor); (2) a preexisting familial history of left-handedness or ambidexterity (genetic factors); and (3) a low level of education (environmental factor). In summary, language disorders within the limits of this particular study were more likely to be observed if the lesion was cortical, if there was a history of left-handedness or ambidexterity, and if the level of education was low. Moreover, the least performing right-brain–damaged subjects (in relative terms) tended to have a lesion that occurred earlier, was larger in size, and affected the retrorolandic region.

In spite of the limitations of this study, the most obvious being the significance of the errors observed and their linguistic nature, the demonstrated influence of anatomical, genetic, and environmental factors has important implications with respect to the formulation of the question of a possible contribution of the right hemisphere to verbal communication. Ideally, this question should be able to specify—either in topographical terms or, more realistically, in terms of neural nets—which neurobiological substrate would be the most susceptible to allow those processes that are necessary for any verbal communication. Furthermore, the question should also be able to specify for which individuals, described in genetic and in environmental terms, this participation is more likely to occur, and to what degree. It is obvious that no straightforward answers to these precise questions are currently available. In fact, the vast majority of the studies discussed and referred to in the following chapters do not touch on these aspects of the question. Rather, they attempt to define, either implicitly or explicitly, the nature of the possible contribution of the right hemisphere, considered in its entirety, with almost total disregard for the genetic and environmental factors capable of influencing this contribution.

Another dimension that has been overlooked in all of the studies addressing the question of a possible contribution of the right hemisphere to verbal communication is the age factor. Indeed, it is believed that the functional lateralization of the brain for language is less well established in young children than in adults, as indicated by the higher incidence of aphasia following right-hemisphere lesions in young children as well as by their much better rate of recovery from aphasia (Alajouanine & Lhermitte, 1965; Basser, 1962; Hécaen, 1976; Teuber, 1975;

Vargha-Khadem, O'Gorman, & Watters, 1985; Woods & Teuber, 1973). However, even in early childhood, verbal communication is affected more by the effects of left-brain damage than by those of right-brain damage (Vargha-Khadem et al., 1985; Woods & Carey, 1979; Rankin et al., 1981). Nonetheless, the contribution of the right hemisphere to verbal communication does seem more likely during early childhood. However, most of the studies postulate—at least implicitly—that the possible contribution of the adult right hemisphere to verbal communication is, for a given individual, the same throughout adulthood and aging. Some authors have proposed models of the functional lateralization of the brain for one aspect of verbal communication, namely language, which challenge this conception.

The best known of these models is that of Jason Brown (1976, 1982; Brown & Jaffe, 1975). Simply stated, this model suggests the existence of both an inter- and intrahemispheric age-related restriction of the methodological basis of the language functions. In other words, during the course of maturation, the left hemisphere and certain poles within the language area become gradually specialized to subtend linguistic functioning. This inter- and intrahemispheric restriction appears to follow a certain hierarchy in terms of linguistic functions, with processes of expression being restricted before those of comprehension and phonetic-syntactic aspects being restricted before semantic ones.

The rationale for this model integrates the idea that the dynamic processes observed during childhood do not completely stop in adulthood but instead are slowed down. However, Brown's main argument in favor of this conception stems from the literature on the changes in aphasia type with age (see Boeglin, Goulet, & Joanette, 1988, and Joanette, Ali-Chérif, Delpuech, Habib, Pélissier, & Poncet, 1983, for reviews). Indeed, it is known that the mean age of Broca's aphasics is approximately ten years younger than that of Wernicke's aphasics. This finding has been replicated in several studies. Moreover, Poncet and Ponzio (1975) suggest that the most prototypical Wernicke's aphasics with logorrhea are mostly found after the age of 70. This finding is not due to an age-related change in lesion distribution (Habib, Ali-Cherif, & Poncet, 1987), despite controversies about this point (Boeglin et al., 1988). Thus Brown, considering that age-related cognitive, memory, and psychosocial changes are not sufficient to account for this result, suggests that the neural substrate underlying the functional organization for language itself is the object of changes with age. In this respect, Brown's model could be considered as one particular application of the dynamic model proposed by Goldberg and Costa (1981).

If Brown's model were to be confirmed, there would certainly be repercussions concerning our expectations as to the possible contribution of the right hemisphere to verbal communication. Indeed, one might expect the possible contribution of the right hemisphere to change with age, even in adulthood, in the sense of a diminution of its contribution. Thus studies of young normal subjects might overestimate the potential of the right hemisphere, whereas studies looking at the effects of right-brain damage in older subjects (e.g., typical mean age in the fifties) might underestimate it. The only study that has looked at possible age-related differences in certain linguistic skills (e.g., word naming, object naming, word comprehension) in a right-brain–damaged population is, however, inconclusive (Joanette, Drolet, & Goulet, 1989). Still, this aspect of the question should not be overlooked in future studies and should be considered along with the anatomical, genetic, and environmental components mentioned earlier.

The Question

As this chapter has shown, the historical antecedents of the question of a possible contribution of the right hemisphere to the maintenance of processes that are required for verbal communication are to be found in its corollary question, that of the contribution of the left hemisphere to verbal communication. However, as time passed, the question of the contribution of the right hemisphere had precursors, initiators, and limits. In spite of its naive and general character, this question has an obvious clinical and theoretical importance, as attested to by the large number of studies that have examined this question since the early 1960s and which are, for the most part, summarized and discussed in this monograph. As one will see, the gradually improved theoretical framework that has been used to examine this question has caused it to shift from language itself to the more general concept of verbal communication. In doing so, the question reflects better the early clinicians' concern about the effects of a right-hemisphere lesion on communication behavior, language included.

Thus the following chapters discuss the question of the possible contribution of the right hemisphere to verbal communication. For some, this question might seem irrelevant and unfounded. However, it has yet to be proven that the sole responsibility for ensuring the integrity of verbal communication lies within and only within the neural nets of the left hemisphere.

2
Methodological and Conceptual Limitations

Our knowledge of the relationship between the right hemisphere and verbal communication and, in particular, the relationship between the right hemisphere and language, is based on the convergence of findings from studies employing different populations and different methods. Each experimental population (e.g., normal, right-brain–damaged, or split-brain subjects) and each experimental method provides evidence that is specific to each population and/or method. The purpose of this chapter is to examine certain conceptual and methodological limitations of the findings in studies of each of the groups just mentioned. Among other things, this chapter discusses the various methodological factors, the underlying models, and the limitations of the interpretations, all of which should be kept in mind every time that a study is discussed in this monograph. The chapter is divided into three sections pertaining to the three most frequently studied populations in the literature: right-brain–damaged, commissurotomized, and normal subjects.

Studies of Brain-Damaged Subjects

Historically, the first experimental investigations of the question of an eventual contribution of the right hemisphere to verbal communication involved right-handed, right-brain–damaged subjects (see Chapter 1). Apart from the usual precautions one has to take with brain-damaged subjects (e.g., etiology, lesion site, lesion size, postonset time, handedness, etc.), these studies were not without a certain number of methodological and conceptual problems that were either specific to this approach or similar to those found in other studies employing different methods and populations.

Neuropsychological and Linguistic Specificity

One of the first questions that arises when documenting the impairment of verbal communication in right-brain–damaged subjects concerns the specificity of this impairment. This question is important if one believes that certain aspects of verbal communication are subtended exclusively by the right hemisphere. It brings to light, among other things, the importance of selecting appropriate normal control subjects who are to be compared with right-brain–damaged subjects. Studies conducted by Francois Boller and his colleagues allow this problem to be posed in concrete terms. To ensure that the language disorder reported in a given study is not simply the result of the presence of a lesion anywhere in the brain, one would have to prove that the disorder is present in right-brain–damaged patients but not in nonaphasic left-brain–damaged patients. As we will see from the different studies presented in this book, the latter population is far from being systematically taken into consideration. In reality, the very idea of having to distinguish between aphasic and nonaphasic left-brain–damaged subjects is not easy to operationalize. For example, Boller and Vignolo (1966) used simple comprehension and naming tasks to distinguish between aphasic and nonaphasic left-brain–damaged subjects. However, when a more complex comprehension task was used, such as the Token Test (De Renzi & Vignolo, 1962), the performance of the nonaphasic left-brain–damaged subjects was inferior to that of right-brain–damaged patients whose performance was similar to that of normal subjects.

Boller (1968) examined the performance of aphasic, nonaphasic left-brain–damaged subjects, and right-brain–damaged subjects on naming and verbal fluency tasks. Performance on the naming tasks (i.e., naming objects and parts of objects) was similar in all three groups. However, the verbal fluency task (i.e., naming as many animal names as possible in one minute) was failed in much the same way by the left-brain–damaged and the right-brain–damaged patients. The results of this study illustrate how important it is to be cautious when interpreting inferior performance of right-brain–damaged patients on verbal tasks. A finding such as this does not necessarily mean that the affected component of verbal communication is exclusively subtended by the right hemisphere. To prove this, an adequate control group is required. It should be noted, however, that quantitatively similar results, such as those of Boller's study, are not enough to reject a specific involvement of the right hemisphere with respect to a given component of verbal communication. As we will see in subsequent chapters, it is possible to uncover specific contributions through the qualitative analysis of results, in spite of their quantitative equivalence.

An equally important issue is that of the purely linguistic specificity of language disorders subsequent to a lesion of the right hemisphere. This problem is illustrated here through a series of studies in which the verbal communication disorders of right-brain–damaged patients were examined in light of concurrent hemineglect and self-denial. Ever since Babinski's (1914) original observations, numerous researchers (e.g., Denny-Brown, Meyer, & Horenstein, 1952; Gainotti, 1972) have described the behavioral indifference of right-brain– damaged patients, which may coexist in varying degrees with a syndrome of hemineglect. These studies cannot be described here. Weinstein was one of the first authors to draw attention to the verbal aspects of behavior in right-brain–damaged patients. Whether dealing with the jargon of Wernicke's aphasics (Weinstein & Puig-Antich, 1974) or the anosognosia of right-brain–damaged patients (Weinstein & Kahn, 1955; Weinstein & Lyerly, 1972), the originality of Weinstein's articles was in their concern with explaining patients' communicative behavior. Weinstein's main objective was not so much to describe the patients' linguistic behavior as it was to apprehend the quality of their communication skills. Using terms identical to those of Eisenson (1962) or Critchley (1962), Weinstein (1963, 1964) drew attention to the particularities of verbal communication in right-brain–damaged patients (or at least some of them): their tendency to use circumlocutions, to make tangential remarks with respect to the theme of discussion, and to use metaphors in describing or denying their handicap. The nature of the spontaneous expression of these patients and that of their verbal contact with their interlocutor led to the suggestion that anosognosia might be responsible for their behaviors (Cutting, 1978; Weinstein & Kahn, 1955; Weinstein & Keller, 1963). Indeed, anosognosia should be considered not only as a perceptual disorder or an impairment in body image, but also as a disorder of one's relationship with the environment.

The view that the verbal communication abilities of the right-brain–damaged patient are indissociable from the patient's attitude toward the disorder is also reflected in the choice of verbal tasks that Weinstein administered to his patients. In addition to the classic tasks of naming familiar objects and naming parts of objects (Weinstein & Keller, 1963), Weinstein (1964) included a naming task consisting of objects associated with the illness (e.g., a syringe, a wheelchair). The patients were also asked to explain metaphoric expressions that involved a part of the body (e.g., to have legs like jelly). We will not go into the details of the different comparisons between aphasics and right-brain–damaged patients since the lack of anatomical evidence and group heterogeneity open such comparisons to criticism. However, it is important that we at least report the results for the ten right-brain–damaged patients examined by Weinstein. Seven of these patients made more than one naming error, which, when compared to the errors of aphasic patients, dealt almost exclusively with the category of objects associated with the illness.

Another interesting finding was that these errors occurred in patients who had either hemineglect or anosognosia. Thus, according to Weinstein (1964), naming disorders, such as literal explanations of idiomatic expressions, are nothing more than the expression of a more global disorder of behavior. Although the examples provided by Weinstein seem caricatural and concern only a subgroup of right-brain–damaged patients (i.e., anosognosics), it remains true nonetheless that an explanation in terms of a behavior disorder should be kept in mind. However, most of the recent studies that have focused on this aspect or that aspect of language tend to overlook this behavioral dimension. These aspects of verbal communication are dealt with in more detail in the chapter on pragmatic aspects (Chapter 7).

Experimental Design and Group Composition

The importance of individual factors requires that subjects from different groups be matched according to the classic variables such as age, sex, sociocultural level, and handedness (Joanette, Lecours, et al., 1983; Meier & Thompson, 1983). The existence of a medical condition, whatever it may be, is also likely to impair test performance. Thus it is preferable that patients who have been hospitalized for reasons other than neurological be included in the control group (Hier & Kaplan, 1980; Klonoff & Kennedy, 1966). The nature of the lesion (e.g., tumor, trauma, vascular, or epileptogenic) and the time since onset must be controlled: any lack of certitude concerning the uniqueness of the lesion, its site, or the stability of the patient's general functional level can result in tenuous or false interpretations (see, for example, Woods & Teuber, 1978, on Basser's 1962 conclusions). As a rule, one should study patients with a well-defined lesion, at a distance from the acute phase. A control group of nonaphasic left-brain–damaged patients should be included to verify the specificity of the disorders observed in the right-brain–damaged patients.

As we will see throughout this monograph, many of the recent studies have involved relatively few patients, and this tendency has become more pronounced as the methodology increases in complexity (e.g., Shapiro & Danly, 1985). These studies are at the crossroads between group studies and multiple single-case studies (Caramazza, 1986). Such studies are questionable from the methodological standpoint as they do not fulfill the necessary conditions of either one of the experimental designs just mentioned. In fact, very few studies of the possible contribution of the right hemisphere to language have employed a single-case or multiple single-case study experimental design. It is possible that this situation reflects the lack of relevant theoretical models sufficiently precise and operational concerning the possible contribution of the right hemisphere. Whatever the reasons might be, the fact remains that studies of right-brain–damaged patients, with occasional exceptions (e.g., Theroux, 1987), have relied on the more or less skillful use of the group study experimental design.

Nature of Associated Disorders

The observation of difficulties on verbal tasks is not enough to prove that a language disorder itself is the cause. It is important to make use of additional information to ensure the validity of the verbal tasks and, above all, to allow the interpretation of the performance recorded.

With respect to methodology, it is obvious that the presence of hemineglect or visuoperceptual disorders requires a careful analysis of the results obtained with visually presented materials. We will see, on the one hand, that the absence of information concerning the visuoperceptive abilities of subjects can interfere with the interpretation of certain findings (e.g., Eisenson, 1962) and, on the other hand, that taking such information into consideration can alter the conclusions of a given study (Bishop & Byng, 1984; Gainotti, Caltagirone, & Miceli, 1983).

The rule then is to always inquire about the possible influence of any visuoperceptual disorders.

With respect to the interpretation of performance, we will be led to consider the significance of correlations between language disorders and a decrease in intellectual efficiency in areas as diverse as lexicosemantic comprehension (Gainotti et al., 1983) and intonation comprehension (Tompkins & Mateer, 1985). Occasionally, we will also examine these associated disorders in their more general aspects with respect to the behavior of right-brain–damaged subjects. At this point, we would like to emphasize two opposing trends: (1) the classic, caricatured description of the right-brain–damaged subject—heminegligent, anosognosic, indifferent to the disorder and to the environment (Babinski, 1914; Weinstein & Kahn, 1955) and (2) the detailed analysis of the right-brain–damaged subject's performance on a particular task such as the comprehension of sentence pairs (Brownell, Potter, Bihrle, & Gardner, 1986).

In the first case, the description of communication behavior is essentially valid for certain "prototypical" patients and emphasizes the emotional component of this behavior (Gainotti, 1972). In the second case, studies of the more complex aspects of language (Gardner, Brownell, Wapner, & Michelow, 1983) tend to disregard the overall behavior of the individual who performed the task, despite the fact that the subject's performance is analyzed in detail. In fact, each time that a study analyzes the disturbance of a given aspect of verbal communication, one should ask to what extent a more general disorder is not contributing to the disturbance in question. The importance of taking into consideration the overall behavior of right-brain–damaged subjects will be discussed in more detail in the chapter on the pragmatic aspects of communication (Chapter 7).

Visual Neglect

In the preceding paragraphs, mention was made of an association between verbal communication disorders and the neglect syndrome. Another reason why the eventual presence of this syndrome should be taken into consideration has to do with methodology. Consider, for example, the Token Test used by Boller and Vignolo (1966) and Boller (1968). This test requires that the subject indicate or manipulate tokens differing in shape, color, and size. One must therefore make certain that the subject does not present a hemineglect that would lead to the consideration of possible solutions only in the right visual hemifield. Unfortunately, in these two studies, the authors made no mention of the spatial distribution of their subjects' errors.

This methodological flaw can help explain some of the contradictory results that have been obtained with the Token Test. For example, Swisher and Sarno (1969) reported a comprehension disorder, whereas Lesser (1974) and Cavalli, de Renzi, Faglioni, and Vitale (1981) reported that the performance of right-brain–damaged subjects was similar to that of normal controls. As we will see (e.g., Gainotti et al., 1983), controlling for hemineglect is of primary importance in the interpretation of findings from studies employing visual materials.

Use of Visual Materials

The use of pictures and drawings poses additional problems. When iconographic material is used, one should refrain from automatically attributing expression or comprehension errors to a language disorder. As Critchley (1962) pointed out, it might be that illustrations are, in themselves, the source of the interpretation difficulties that are observed in right-brain–damaged patients. The results of a study conducted by Archibald and Wepman (1968), in which illustrations were used, exemplifies this criticism. After having submitted 39 left-brain–damaged and 22 right-brain–damaged patients to a series of language tests, Archibald and Wepman reported the results of only 8 of the 22 right-brain–damaged patients, that is, those who displayed the most significant disorders.

The subjects were compared with a group of six syntactic aphasic patients as well as with a group of nonaphasic left-brain–damaged patients. Some of the tests that were used in this study included oral naming of pictures, reading, comprehension (e.g., pointing to pictures of words that were spoken or read), and making up a story with four pictures. The results revealed that naming and reading errors occurred more frequently in right-brain–damaged patients than in nonaphasic left-brain–damaged patients, but less so than in aphasics. However, errors in pointing to pictures were more frequent in right-brain–damaged patients than in aphasics. This tends to suggest that some of the right-brain–damaged patients' errors were in fact secondary to perceptual disorders and/or a disorder of visual exploration. The authors provided no evidence to the effect that the pictures themselves were not the cause of any specific problems for the subjects. As we will see, some authors have interpreted the erroneous responses of right-brain–damaged patients as a difficulty in picture discrimination (Bishop & Byng, 1984).

Visuospatial Disorders

Hier and Kaplan (1980) assessed verbal and visuospatial abilities in order to uncover any eventual relationship between the performance on each of these two types of tasks. No visual materials were used in the verbal tasks, which included the Wechsler vocabulary subtest, a sentence comprehension task with a simple yes-or-no response, and a task on the use of metaphoric expressions in explaining various proverbs (e.g., "Don't cry over spilt milk"). The selection of brain-damaged subjects and normal controls was based on strict criteria. A comparison was made between 34 right-handed patients, examined within the first two months following a stroke and 16 patients who had been hospitalized equally long for nonneurological illnesses. Both groups of subjects performed equally well on the Wechsler vocabulary subtest. However, on the sentence comprehension task, the performance of 22 of the 34 brain-damaged patients was significantly less than that of the control subjects. Furthermore, 21 of these same brain-damaged patients exhibited inferior performance on the proverb explanation task. With respect to the sentence comprehension task, passive sentences (e.g., "Mary was phoned by Fran. Was Fran phoned?") as well as sentences implying spatial relationships (e.g., "The elephant sat on the mouse. Was the mouse on top?") were failed, whereas sentences involving a comparison (e.g., "Are trains faster than airplanes?") or a temporal relationship (e.g., "Does lunch come before dinner?") were performed more accurately.

In this study, memory deficits could not account for the inferior performance of the brain-damaged patients since the sentences were repeated at the patient's request. It is not entirely impossible that attention deficits were to blame, although in this study, as well as in others that followed, no attempts were made to control for specific (e.g., visual) or general attentional factors. Notwithstanding the fact that the performance of the brain-damaged patients on the Wechsler vocabulary subtest was similar to that of the normal controls, we know nothing of their intellectual efficiency. Hier and Kaplan (1980), however, attached much importance to the significant correlation between the scores on the two verbal tasks and those on the three visuospatial tasks (i.e., copy of Rey's figure, Kohs blocks, and identification of hidden figures). They concluded that a mental imagery problem, such as the one hypothesized by Caramazza, Gordon, Zurif, and De Luca (1976), could partly account for the inferior performance of the brain-damaged patients. Although this is a likely hypothesis in the case of sentences whose content involves a spatial relationship, it remains to be seen whether it can be applied to the comprehension of linguistic information in general.

Decline in Intellectual Efficiency

The problems that arise from the possible effects of a nonspecific decline in intellectual efficiency are illustrated in the Archibald and Wepman's (1968) study described earlier. Indeed, the intellectual efficiency of the 8 language-impaired right-brain–damaged patients was significantly less than that of the 14 nonaphasic right-brain–damaged patients. This result can be interpreted in two entirely different ways. On the one hand, the presence of a right-hemisphere lesion leads to a decline in the overall efficiency of the subject. A correlation between the magnitude of this deficiency and performance on the verbal tasks would tend to suggest that poor performance is more the result of a global deficit than the expression of a specific disorder of the right hemisphere's contribution to language. On the other hand, poor performance on the efficiency tasks and the verbal tasks might have a common cause. To assess the intellectual efficiency of their subjects Archibald and Wepman used the Raven Progressive Matrices Test, a test that requires the subject to select one of several visual patterns that will logically complete a series. They concluded that inferior performance on the language tasks was due to the presence of general mental deterioration. However, if visual exploration is impaired in a right-brain–damaged patient, either through hemineglect or a visuoperceptual disorder, it follows that the patient will fail the Raven Test, as well as any language test that uses visual materials.

In summary, Archibald and Wepman's (1968) intepretation in favor of nonspecific mental deterioration cannot be taken for granted unless evidence is provided concerning the absence of any difficulties with the visually presented material. As we will see (Chapter 4), the answer to this dual issue of visual processing and general intellectual efficiency requires a certain amount of methodological control (Gainotti et al., 1983).

Studies of Split-Brain Subjects

Several hypotheses concerning the possible contribution of the right hemisphere to verbal communication have emerged from the in-depth study of commissurotomized, or split-brain, subjects. Of particular interest is the performance of these subjects on tasks dealing with certain aspects of linguistic function. The relevance of such studies for the understanding of hemispheric functioning has led to so many developments, comments, and hypotheses, that, more often than not, the particular nature of this population has been forgotten, thereby leading to excessive extrapolations in some cases. At this point it seems essential to specify the characteristics of these subjects and the limitations of the evidence that has been collected since 1970.

Characteristics of the Population

Surgical sectioning of the corpus callosum has been performed on patients suffering from intractable epilepsy. By definition, these patients have sustained lesions in either one or both hemispheres early in life and supposedly have undergone a reorganization of their hemispheric capacities. In particular, an epileptogenic lesion of the left hemisphere could result in the transfer of certain linguistic capacities to the right hemisphere (or the absence of restriction to the sole left hemisphere). The history of each of these patients, the nature of the epileptogenic lesions, their associated disorders and intellectual efficiency, and the nature of the surgery (i.e., complete versus partial section of the corpus callosum) all suggest that these patients (approximately 40 in all) cannot be considered as a single, homogeneous group (Myers, 1984; Whitaker & Ojemann, 1977). For the same reasons, very few of these patients have been extensively examined. Finally, it is not known if the degree of functional lateralization of the brain for language to the left hemisphere is similar to that in normal, nonsplit individuals.

Commissurotomy and the Transfer of Information

So as not to consider commissurotomy as being a total, complete absence of transfer of information, several remarks are warranted. The first point has to do with methodology. These veteran testees are able to use, at least on certain tasks, facilitating, or cross-cueing strategies (e.g., eye or head movements in response to vertically arranged stimuli, or subvocal evocation of the possible solutions) (Beaumont, 1982b; Gazzaniga, 1970; Gazzaniga & Hillyard, 1971). These strategies are not easily controlled and should be taken into consideration when interpreting the results of these patients (see Zaidel & Peters, 1981, patient L.B.).

The second point concerns the role of the spared anterior commissure (Sperry, 1982). The nature of the information that supposedly passes through this commissure is a controversial issue (Myers, 1984). The sparing of this commissure could be favorable to the development of linguistic capacities within the right hemisphere (McKeever, Sullivan, Ferguson, & Rayport, 1982).

The third and final point, and by far the most important, is the possibility of subcortical transfer of information. In a number of studies, the findings from both verbal and nonverbal tasks, presented under particular experimental conditions (e.g., elimination of cross-cueing strategies and allowance of longer response times), suggest that the left hemisphere has access to information concerning stimuli that have been presented solely to the right hemisphere. This information is sufficiently informative to allow for correct naming or for comparison with stimuli presented to the left hemisphere (Johnson, 1984; Myers & Sperry, 1985; Sergent, 1983). The information that passes through the subcortical pathways appears to be partial. However, even if the stimulus is not readily identifiable, the information concerning its context or its meaning is sufficient for it to be recognized by the left hemisphere (Gazzaniga, 1983a; Sidtis & Gazzaniga, 1983; Sperry, Zaidel, & Zaidel, 1979).

Interpretation of Findings

The number of commissurotomized subjects whose right hemisphere has acctually been found to demonstrate linguistic capacity varies between five and eight (Beaumont, 1982b; Gazzaniga, 1983a; Gazzaniga & Baynes, 1986; Zaidel, 1983b). In certain patients (e.g., P.S. and V.P.) this capacity is secondary to the presence of a massive epileptogenic lesion in the left hemisphere. In other patients (e.g., L.B. and N.G.) this argument is less likely (Zaidel, 1983a; Zaidel & Schweiger, 1984), though it is debatable (Gazzaniga, 1983a; Sidtis, Volpe, Wilson, Rayport, & Gazzaniga, 1981).

In line with these facts, one should not conclude that these patients are representative of a somewhat exceptional minority (Gazzaniga, 1983a), since establishing a within-group frequency (e.g., 5 patients out of a total of 44; Gazzaniga, 1983a) presupposes homogeneity of the original group, which is far from being the case here. In reality, few patients fulfill all of the conditions to allow testing of the linguistic capacity of the right hemisphere (Myers, 1984). As a result, any estimate in terms of frequency has little validity. Outside of any controversy concerning one patient or another, one should simply keep in mind that the published facts concern a small number of patients (Gazzaniga, 1983a, 1983b; Levy, 1983b; Myers, 1984; Zaidel, 1983a). What are the implications of these facts with respect to our knowledge of the contribution of the right hemisphere to language in the normal subject? Two types of answers to this question can be reported. The first is that of Millar and Whitaker (1983), who believe that these facts are of little interest since one is dealing with particular subjects in whom the right hemisphere has been disconnected from the left hemisphere and thus is not comparable to the normal right hemisphere. The second is that of Gazzaniga (1983a, 1983b), who used this minority of subjects within the general population of commissurotomized patients to demonstrate that the capacities of the right hemisphere are limited and observable only in a small number of individuals. The former notion rejects any comparison with normal functioning, whereas the latter proposes an optimistic view of the situation based on a supposedly homogeneous population. Between these two extremes, it seems to us that the contribution of the study of these patients can be considered in one of two ways.

First, they help determine the upper limit of the potential capacities of the disconnected right hemisphere—upper limit insofar as the postlesion plasticity can result in the right hemisphere being better equipped than that of a normal subject, and potential abilities insofar as one is not dealing with the contribution of a right hemisphere that is connected to the left hemisphere. The information provided is therefore of a negative value; if certain tasks are not performed by a disconnected right hemisphere, it is conceivable that the neural organization of the normal right hemisphere does not support the processing of such tasks.

Second, the capacities demonstrated in these patients provide information about the nature of the processes that are likely to occur in the normal right hemisphere. The findings have no quantitative value. Rather, it is the nature of the process uncovered that is informative (Rabinowicz & Moscovitch, 1984). This concerns potential processes, which are not necessarily those to which the right hemisphere contributes when it is normally connected to the left hemisphere. However, if these processes are congruent with those suggested by studies of other populations, brain-damaged or not, then it is admissible, on the basis of the convergence of these findings, that these processes are involved in the functioning of the normal right hemisphere. It is from this point of view that studies with commissurotomized patients have become the stimulant for devising similar experiments with normal subjects (Beaumont, 1982b).

In summary, each time that a study based on the performance of split-brain subjects is reported in this monograph, one should (1) remember that we are dealing with particular individuals and take into consideration what their right hemisphere *cannot* do, (2) ask if the performance of the right hemisphere can be supported by studies of normal subjects or of other populations, and (3) remember that the findings, like those obtained in studies of normals, reflect the potential of the right hemisphere but are not necessarily representative of the actual contribution of the right hemisphere to verbal communication.

Studies of Normal Subjects

In this section we examine the methodological and conceptual problems associated with the study of the possible contribution of the right hemisphere to verbal communication—most frequently evaluated through the ability to process words—in normal right-handed adults, that is, subjects who show no signs indicative of damage to the central nervous system.

Individual Factors

Various individual and cultural factors can have an effect on the presence of lateral differences. No other chapter in this book specifically deals with factors such as sex, age, laterality, sociocultural level, general knowledge (e.g., vocabulary size, knowledge of the rules of word formation), previous experience, and second-language use, all of which are liable to influence a subject's response strategies (Bradshaw & Gates, 1978; Dillon & Smeck, 1983; Hiscock & MacKay, 1985; Kimura & Harshman, 1984; Levy, 1983c; McGlone, 1980; Perfetti, 1983; Segalowitz & Bryden, 1983; Vaid, 1983). Nor will the influence of writing be dealt with specifically; for example, we do not consider languages that are written from right to left (Bertelson, 1972) or languages that use ideograms (Paradis, Hagiwara, & Hildebrandt, 1985). Finally, it should be noted that most studies of normal subjects have focused on university undergraduate students. The results of these subjects should not be considered as representative of the general population of normal adult subjects.

Questions to Be Answered

Any behavioral measure is the product of a series of interacting strategies elaborated either within each hemisphere or between the two hemispheres. Such strategies make it difficult to interpret a given measure as reflecting the specific contribution of a single hemisphere. Within such a context, any experimental investigation should control the three types of factors that influence the lateralization of the response: those related to the stimulus, to the task, and to the response.

The nature and the conditions of stimulus presentation must be controlled since it is important to specify the nature of the information that must be extracted from the stimulus in order to solve a given task. Depending on the nature of this information, one hemisphere is more likely to be activated than the other (Sergent & Hellige, 1986).

Successful performance on a given task implies several possible strategies that must be taken into account. The various levels of processing involved within each of these strategies must also be taken into account, since the levels may be of varying complexity and may occur either simultaneously or successively (Bertelson, 1982; Rabinowicz & Moscovitch, 1984).

One also needs to know if the result recorded is due to the processing of the stimulus or if it is the consequence of response planning. This issue can lead to some serious problems. On the one hand, it is possible that response planning and stimulus processing are in direct competition with each other (Green, 1984). One the other hand, the transfer of information via the corpus callosum, needed to activate a manual response, for example, might occur too rapidly for it to be detected in a classic reaction time experiment (Millar & Whitaker, 1983; Zaidel, 1986a).

If indeed there are some contradictions between the numerous studies of normal subjects, it could be that the various methodological issues mentioned here are far from being resolved and that the techniques currently in use are restricted in their evaluation of the nature of cerebral information processing in real time.

Methods

Two methods have been consistently used to demonstrate lateral differences in normal subjects: divided visual field studies using tachistoscopic presentations and the dichotic listening paradigm.

Tachistoscopic Experiments

Tachistoscopic presentations are based on the anatomical characteristics of the visual pathways that transport information from one hemifield to the contralateral hemisphere. The duration of the stimulus presentation must be extremely brief—tachistoscopic—in order to remain below the minimal amount of time that is required to bring the stimulus into central vision by way of microsaccadic eye movements (less than 200 ms, according to Young, 1982). The various aspects of this technique have been reviewed elsewhere (Beaumont, 1982a, 1983; Bryden, 1982; Hellige & Sergent, 1986; Sergent & Hellige, 1986; Young, 1982). The following factors shed some light on why artifacts are possible and why interstudy comparisons can be problematic:

1. Nature of the stimulus (e.g., letters, words, or drawings), its orientation (e.g., horizontal or vertical presentations or words), and its size.
2. The conditions of stimulus presentation: luminance, contrast, spatial frequency, type of screen, degree of eccentricity, duration, use of masking, control of the fixation point, and uni- or bilateral presentation.
3. The nature of the task: naming, lexical decision, categorization, or matching.
4. The nature of the response: vocal or manual (e.g., one or two hands, number of response buttons). In the case of bilateral presentations, one must consider the order in which the responses are to be made. To avoid any attentional biases, this order should be determined by the experimenter, though it should not create a concurrent task (Schmuller & Goodman, 1979).
5. The type of measure used: total number of correct responses or reaction time measures. The ideal stimulus duration should not result in more than a few errors. Thus a stimulus duration that results in no errors at all is unlikely to reveal any differences. Conversely, a high number of errors suggests that the subject has adopted a compromise between accuracy and speed (Beaumont, 1982a). Furthermore, the statistical analysis of the data, either in terms of accuracy (Loftus, 1978) or in terms of reaction time (Birkett, 1977; Levy, 1983b), creates its own set of problems with respect to the interpretation of the results.

Dichotic Listening Experiments

This method consists in the simultaneous presentation of different stimuli to each of the two ears. The subject's task is to identify the stimuli presented. An auditory stimulus is projected to the auditory cortex of both the ipsilateral and the contralateral hemispheres. Using verbal stimuli, Kimura (1961) demonstrated the presence of an advantage of the right ear over the left ear in right-handed subjects: stimuli presented to the right ear were identified more readily than those presented to the left ear. Kimura attributed this finding to the superiority of the contralateral over the ipsilateral auditory pathway that exists within the context of dichotic competition. Therefore, because of the superiority of the left hemisphere for language, verbal stimuli presented to the right ear are processed preferentially. Since then, numerous studies have shown that the results obtained with this technique are far from being a simple reflection of the functional asymmetry of the brain. On the one hand, there are several arguments against Kimura's hypothesis, in particular that of ipsilateral suppression (Geffen & Quinn, 1984). On the other hand, it is a well-established fact that numerous methodological factors have an influence on dichotic listening performance (Bryden, 1982). These factors include:

1. The preparation of dichotic listening tapes, which must take into consideration several acoustical parameters (e.g., intensity, synchronization, and signal-to-noise ratio) (Porter & Hughes, 1983).
2. The nature of the stimulus (e.g., consonant–vowel syllables, words, digits, sentences, and intonation patterns).
3. Memory factors involved in the recall of the stimuli.
4. Shifts in attention strategies (Geffen & Quinn, 1984).
5. The nature of the task (i.e., oral or written; recall or recognition) (Blumstein & Cooper, 1974).
6. The nature of the response (i.e., one ear or two) and the order of response (Bryden, Munhall, & Allard, 1983).
7. Methods used to calculate between-ear differences in performance (Bradshaw, Burden, & Nettleton, 1986).

More detailed reviews of these various factors can be found in Beaumont (1983), Boeglin (1985), Bradshaw et al. (1986), and Bryden (1982).

Interpreting Lateral Differences

Even if it was possible to eliminate all of the methodological artifacts just listed, the problem of interpreting the results would still exist. In the next section we attempt to summarize the various existing models thus showing the reader why no dichotomous generalizations are found in this book. We also want to illustrate the fact that the interpretations are dependent on the questions that are raised as well as on the nature of the methodology that is used.

Anatomical Models of Direct Access or Callosal Relay

The interpretation of the lateral differences observed following tachistoscopic or dichotic presentations can first be considered from a structural, anatomical point of view (Bertelson, 1982; Searleman, 1983; Zaidel, 1983a, 1986a). Two possibilities can be considered. On the one hand, each hemisphere is competent to solve the task, though one is more efficient or faster than the other. This is the so-called direct-access model. On the other hand, only one hemisphere is competent and the stimuli presented to the other hemisphere must pass through the corpus callosum in order to be processed by the competent hemisphere. This is the so-called callosal relay model (Zaidel, 1986a). By using an adequate methodology, which avoids inducing hemispheric activation through the systematic or anticipated use of the response hand (Hellige & Sergent, 1986; Zaidel, 1986a), it is theoretically possible to specify the criteria to be met in favor of the direct-access or of the callosal relay way of considering things (Zaidel, 1983a, 1986a).

For example, the observation of a hand–hemifield interaction supports the direct-access interpretation. In the case of a stimulus presented to the left visual half-field the response is faster with the left hand, whereas in the case of a stimulus presented to the right visual half-field the response of the right hand is faster. The superiority of the hand ipsilateral to the field of presentation indicates that information can be processed by the same hemisphere that receives the stimulus and programs the response. Another piece of evidence that is favorable to the direct-access interpretation is the existence of a lateral difference dependent on the nature of the stimulus. For example, the results of a lexical decision task revealed a lateral difference for abstract words but not for concrete words (Zaidel, 1983a, 1986a). Since the two types of stimuli shared similar visuoperceptual features, this could not account for the differences found. Rather, the difference between the two hemifields suggests that there is a direct access for concrete words, with each hemisphere being competent for this type of word but not for abstract words.

Conversely, the demonstration of an effect of the response hand (e.g., the right hand) and an effect of the hemifield (e.g., the right hemifield), in the absence of any hand–hemifield interaction, supports the callosal relay interpretation. In this case, stimuli presented in the right visual hemifield are projected directly toward the left hemisphere, whereas stimuli presented in the left visual hemifield must be relayed from the right hemisphere to the left hemisphere. Since only the left hemisphere is competent in this case, it follows that the responses programmed by the left hemisphere (right hand) are always faster than those programmed by the right hemisphere (left hand), which has to rely on information coming from the left hemisphere.

Although Zaidel (1983a, 1986a) has been able to find tasks to illustrate each of these models, the fact remains that this concept is essentially one of theoretical value. Zaidel himself admits that "direct access and callosal relay may be thought of as representing limit cases of a theoretical continuum, where the two hemispheres interact to greater or lesser extents during task performance" (1983a, p. 115). Indeed, the reasoning on which each of these models is based is that the response involves but a single level of processing. However, if one admits that (1) the tasks currently in use involve several different levels of processing (simultaneous or successive), (2) these levels of processing create different problems for each hemisphere, and (3) depending on the level of processing under consideration, superiority may vary from one hemisphere to the other, then one should not be surprised that few studies have gathered sufficient criteria to comply with either of these models (Bertelson, 1982).

Dynamic Models

Another model, which was already present in Kinsbourne's (1975) work, emphasizes the dynamic nature of inter- and intrahemispheric relationships (Hellige, Cox, & Litrac, 1979; Hellige & Sergent, 1986). From this point of view, lateral differences are not static but instead depend on the nature of the task and, obviously, on hemispheric differences per se. Kinsbourne (1975) emphasized the fact that hemispheric activation is the determining factor in directing attention toward one hemifield or the other. For example, a task that involves verbal proccessing (e.g., an oral response) will activate the left hemisphere and result in preferred attention for the right visual hemifield. It is likely then that the lateral difference observed is more the result of preferred attention than the reflection of a superiority of processing capacities.

Although Kinsbourne's model cannot explain, on its own, the lateral differences observed (Bertelson, 1982; Zaidel, 1986a), it is nevertheless important for two reasons. The first is methodological, that is, it incites researchers to measure the influence of these attentional factors (Larmande & Cambier, 1981) and to eliminate the artifacts caused by the attentional bias (Geffen & Quinn, 1984). The second is theoretical, that is, the model is not restricted to the lateral distribution of attention but, from a more general point of view, concerns the different processes that involve one or both hemispheres. The execution of a given process by one hemisphere is likely to either activate the same hemisphere for the resolution of another process or compete with this process, the saturation of intrahemispheric capacity thereby resulting in the transfer of activity to the contralateral hemisphere. For example, it has been shown that in identification tasks of tachistoscopically presented letters and shapes, the advantage shifts from the right to the left hemisphere when the number of words to be recalled at the same time increases from two to six (Hellige & Cox, 1976; Hellige et al., 1979).

Complexity of Interhemispheric Differences

The anatomical and dynamic models should not be seen as exclusive. Indeed, it is reasonable to assume that identical processes are activated in each hemisphere (direct access), that one hemisphere has the exclusivity of a particular process (callosal relay), and that the distribution of processes between the two hemispheres or within a single hemisphere is a function of the task at hand (dynamic model).

The methodological issues discussed here, as well as the issues involved in the interpretation of the lateral differences that are found, are incompatible with a dichotomous conception of hemispheric function, which is a rather simple form of modeling brought about by investigations of commissurotomized patients (Ledoux, 1983, 1984). These dichotomies have been based either on the nature of the stimulus (i.e., verbal/nonverbal) or on the level of general principles of functioning that differentiate the left hemisphere from the right hemisphere in terms of analytic versus holistic processing (Bever, 1975; Bradshaw & Nettleton, 1983), local versus global processing (Martin, 1979; Semmes, 1968), and serial versus parallel processing (Cohen, 1973). These concepts, formulated a posteriori as conclusions to a number of experiments,[1] lack the requisite precision for use as experimental hypotheses (Bertelson, 1981; Hellige & Sergent, 1986; Marshall, 1981) and cannot explain the contradictory findings that have emerged from studies of normal subjects. The concepts are insufficient to explain the "multidimensional" nature of intra- and interhemispheric relationships.

The Potential of the Right Hemisphere

Finally, it is important to underline the nature of the findings that have emerged from studies of normal subjects. We are dealing with specific tasks (e.g., lexical decision task) presented under particular experimental conditions. These experiments inform as to the *potential* of the right hemisphere (e.g., the comprehension of concrete words), but this does not necessarily mean that these processes are actually used by the normal subject (e.g., when reading a text that contains concrete words) whose right hemisphere is connected to the left.

Two Hemispheres but One Brain

The reservations directed to the interpretation of the literature concerning commissurotomized patients, as well as the remarks concerning studies of normal subjects, should be enough to caution the reader against any inappropriate interpretation of lateral differences. The interest raised by the study of commissurotomized patients and the presence of lateral differences in normals should not lead to the assumption that hemispheric asymmetry is the juxtaposition of two independent hemispheres. Our goal is to acquire knowledge about the brain whose two hemispheres interact with each other to perform complex tasks such as carrying on a conversation or understanding a written text. The demonstration of lateral differences informs us as to the potential of the right hemisphere. However, as mentioned before, this information should not be confused with the actual contribution of the right hemisphere to verbal communication. The study of right-brain–damaged patients enables one to assess this contribution, albeit indirectly.

To illustrate how the nature of the findings collected to date differs depending on the population being studied, let us look at an example. On the one hand, evidence from studies of normal and commissurotomized subjects supports the potential of

[1]An exception to this is the study by Drews (1984).

the right hemisphere to process concrete, frequent words. On the other hand, evidence from studies of right-brain-damaged subjects has shown that they have difficulties understanding abstract words or the nonliteral aspects of language (Critchley, 1962; Eisenson, 1962). Reported as such, these two types of findings may seem contradictory; above all, they differ in their nature. This monograph presents a critical review of the facts concerning *both* the potential of the right hemisphere and the actual use of this potential. We will see what convergences and questions arise from the comparison of these two types of findings.

3
Aphasia and the Right Hemisphere

Comment se fait-il donc que l'individu rendu aphémique par la destruction partielle ou totale de la troisième circonvolution frontale gauche, n'apprenne pas à parler avec l'hémisphère droit? (Broca, 1865, p. 18)[1]

Broca's (1865) discovery of a particular area of the frontal lobe—the foot of the third frontal convolution—proved to be of great importance with respect to the control of articulated language. At the same time, it has raised the problem of the neural basis of recovery from aphasia when, in the best of cases, articulated language is regained. Two distinct explanations have been offered to account for the rapid recovery of expression following a lesion of this area. The first explanation emphasizes the role of the intact areas of the left cerebral cortex: "... c'est sur l'intégrité relative de telle ou telle partie de la zone du langage ou de la corticalité adjacente et par laquelle la lésion peut jusqu'à un certain degré être compensée, que doit porter actuellement l'étude des lésions de l'aphasie" (Dejerine, 1914, p. 113).[2] The second explanation, advocated by Von Mayendorff (1911) and Henschen (1926), emphasizes the role of the homologous area of the right hemisphere: "... the right third frontal convolution acts as a substitute in case of destruction of the left third frontal. If both the left and the right third frontal gyri are destroyed speech disappears completely and forever" (Henschen, 1926, p. 111). However, a review of several cases involving bilateral lesions of the foot of the third frontal convolution revealed that, contrary to Henschen's assertion, oral expression was not irremediably affected (Levine & Mohr, 1979).

A possible compensatory role of the right hemisphere was suggested following lesions to other structures of the language area by Henschen (1926) and Nielsen (1944). According to these authors, the residual language production and comprehension capacities of aphasics with lesions of the foot of the left third frontal convolution or of the left temporal lobe are attributable to the intact right hemi-

[1]"Why is it that an individual who has become aphemic following the partial or total destruction of the left third frontal convolution does not learn to speak with the right hemisphere?" (author's translation).

[2]"... the study of the lesions causing aphasia should deal with the relative integrity of this part or that part of the language zone, or of the adjacent cortical areas, and by which the lesion can be compensated for, at least to a certain degree" (author's translation).

sphere. For example, Nielsen emphasized the role of the right temporal lobe in explaining the comprehension of simple orders in a patient with Wernicke's aphasia (ischemia of the left first temporal convolution): ". . . he understood all of the above requests with the right temporal lobe" (p. 73, observation 6). However, before concluding that ". . . what a patient can do with language after a lesion of the major side depends on the inherent ability of the right hemisphere" (p. 75), one must be able to prove that the intact areas of the left hemisphere are incapable of taking over all, or even part of the residual language capacities (Goldstein, 1948). It is obvious that the observations of the right hemisphere's partisans neither answer this question nor authorize Henschen to draw up a list of the capacities and limitations of the right hemisphere solely on the basis of the performance observed in aphasics.

The debate concerning the involvement of the right cerebral hemisphere in the residual language of aphasic patients has progressed very little. For example, the contrasting views of Landis, Regard, Graves, and Goodglass (1983) and Marshall and Patterson (1983) concerning the eventual role of the right hemisphere in the development of semantic paralexias (see Chapter 5) are somewhat similar to those of the authors previously mentioned. Researchers have yet to develop adequate research methods that would allow them to confirm, reject, or quantify this participation of the right hemisphere. Much of the available evidence supporting participation of the right hemisphere in aphasia has been indirect. This evidence has emerged from three types of observations: (1) those in which compensation by the right hemisphere is highly probable since the lesion of the left hemisphere is massive and affects the entire language area; (2) those in which this participation is inevitable, as in the rare cases of left hemispherectomy; and (3) those in which the contribution of the right hemisphere is inferred indirectly from the aggravation of aphasic symptoms following either a lesion of the right hemisphere or injecting the right hemisphere with sodium amytal. The evidence is limited not only in terms of the small number of actual observations but also in terms of the type of investigation that is warranted by the clinical status of the patient (e.g., hemispherectomy) or the technique used (e.g., sodium amytal). It should be noted that the hypotheses concerning the possible role of the right hemisphere in the production of semantic paralexias, within the context of certain reading disorders (e.g., deep dyslexia), will be examined in Chapter 5.

Destruction of the Language Area

It seems reasonable to assume that the right hemisphere is indeed involved in the residual language of aphasics when the left-hemisphere lesion is massive and involves all of the language zone. The first case described by Landis, Cummings, and Benson (1980; see also Cummings, Benson, Walsh, & Levine, 1979) was that of a 54-year old man, victim of an embolic infarct of the entire sylvian region. Three years later, his initial severe global aphasia had changed considerably. Spontaneous expression had improved, though it was reduced to the use of short sentences and isolated words. The improvement of his comprehension skills enabled the patient to designate objects as well as parts of the body. The second case described by Landis et al. shows that improvement attributed to the right hemisphere can take place over a relatively long period of time. The patient was a 32-year-old male with global aphasia brought on by an ischemia in the territory of the middle cerebral artery following a traumatic occlusion of the carotid artery. Five years after the insult, his spontaneous expression was limited to stereotyped expression, word repetition was possible, and comprehension was limited to simple orders. Nine years after the lesion, his spontaneous vocabulary had improved, though it consisted only of substantives, whereas his comprehension allowed him to designate four consecutive objects. In both cases, the destruction of the classical language area suggests that the right hemisphere was responsible for the evolution observed. The fact that the improvement of comprehension skills exceeded that of expression skills in both patients is compatible with what is known about the linguistic potential of the right hemisphere. It should be noted, however, that in both patients the left thalamus was intact, though the exact relation between this fact and the progress observed is difficult to interpret.

The limitations of the right hemisphere's linguistic potential are illustrated in a case described by Cambier, Elghozi, Signoret, and Henin (1983, case 2). The patient had an ischemia in the territory of the middle cerebral artery as well as in the territory supplied by the anterior cerebral artery. Spontaneous expression was impossible, but upon prompting by the examiner the patient was capable of naming the days of the week, offering a series of numbers, reciting a fable, or completing a sentence. The restriction of his expression brings to mind the hypothesis of Hughlings-Jackson (1915) to the effect that the right hemisphere is involved in the automatic use of words. As we will see, these features of expressive language are similar to those observed in adult cases of left hemispherectomy.

Hécaen attributed the progress of seven patients with pure alexia subsequent to a left occipital lobectomy to the possible involvement of the right occipital lobe (Hécaen, 1976; Jeannerod & Hécaen, 1979). This favorable evolution is in marked contrast to the usually poor recovery from alexia secondary to an ischemia of the territory supplied by the posterior cerebral artery. According to Hécaen, the right hemisphere's capacity for reading becomes accessible following the removal of the left occipital lobe. However, this contribution can be inhibited by the presence of an ischemic lesion. As we will see in Chapter 5, this inhibitory effect of the lesioned left lobe on the right is implicitly part of the model proposed by Coltheart (1983) in cases of alexia without agraphia.

Left Hemidecortication in Adulthood

Occasionally there are medical reasons for patients to undergo total removal of a cerebral hemisphere. At one time, right hemispherectomies, or rather hemidecortications, were, for obvious reasons, performed more often than left hemispherectomies (Dandy, 1928). However, a certain number of cases of *left* hemispherectomy, or hemidecortication, have been reported, though the consequences of this intervention are such that it is seldom undertaken. Of the few reports that have been published, most have provided incomplete information given the clinical status of the patients and their rather short period of postsurgery survival. The first published case was that of a 43-year-old woman operated on by Zollinger (1935). Following the surgery, her expression was reduced to a few words (e.g., sleep, yes), which were used inappropriately. However, because the patient survived for only 17 days, Zollinger was unable to verify his impression that her vocabulary was gradually improving with time. Crockett and Estridge (1951) reported the case of a 37-year-old man. Two weeks after the surgery, he could pronounce a few expressions (e.g., "Put me back to bed"). Over the next month, he continued to improve before his condition suddenly worsened. He was able to copy certain letters but could not read them. French et al. (1955) provided few details concerning their patient, a 38-year-old male who underwent hemidecortication 12 years after surgery for a left parietal glioma. His verbal expression was limited while his comprehension appeared to be partially preserved.[3]

The most frequently mentioned case is that of Smith and Burklund (Burklund & Smith, 1977; Smith, 1966). They described the evolution of language behavior in a female patient over a period of two years following her hemispherectomy. It should be noted, however, that eight months beforehand, the patient (E.C.) had undergone an excision of the left frontal and parietal ascending convolutions, which could have contributed to a certain functional reorganization. Following the hemispherectomy, her verbal expression was restricted to a few words and automatic expressions, with the occasional formulation of three- and four-word sentences. However, her oral comprehension improved considerably. Six months after the surgery, E.C. was capable of selecting the image corresponding to a spoken word from an array of four choices (Smith, 1966). Her reading ability and written comprehension were very limited. Burklund and Smith (1977) described a somewhat similar evolution in a 41-year-old man, over a period of nine months, before his condition suddenly deteriorated.

Despite the interest of these observations, it is difficult to draw any strong conclusions on the basis of such a small number cases. They do seem to suggest, however, that the intact right hemisphere might be responsible for the recovery of comprehension skills (Searleman, 1977; Zangwill, 1967). In fact, the oral comprehension skills of the two patients described by Burklund and Smith (1977) are very similar to those of the right hemisphere in the commissurotomized subjects L.B. and N.G. (Lambert, 1982a).

Aggravation of Aphasic Semiology Following a Right-Hemisphere Lesion

The aggravation of aphasia following a right-hemisphere lesion has been known for a long time. Judging by Gowers several such cases had already been described in the late nineteenth century:

[3]". . . he could make simple wants known, and had a fair degree of comprehension" (French et al., 1955, p. 162).

> Loss of speech due to the permanent destruction of the speech region in the left hemisphere has been recovered from, and that this recovery was due to supplemental action of the corresponding right hemisphere is proved by the fact that in some cases, speech has again been lost when a fresh lesion occurred in this part of the right hemisphere. (1887, pp. 131–132)

Kuttner's 1930 case study reported by Nielsen (1944) was that of a 54-year-old woman with rapidly amending motor aphasia up until the sudden occurrence, one year postonset, of a global aphasia accompanied by a left hemiplegia. The anatomopathological examination revealed an infarct softening of the left putamen and insula. Furthermore, the right hemisphere had been the site of two recent lesions, one affecting the lenticular nucleus and the other, the foot of the third frontal convolution (Broca's area).

More recently, Cambier et al. (1983) reported the case of a female patient with severe aphasia: her expression was limited to the use of automatic expressions and her comprehension to that of simple commands. Two years after the onset of the lesion, the patient had made considerable progress. Her verbal expression now included the use of substantives. Her picture-naming performance revealed semantic paraphasias, but she was able to repeat three-word sentences. The patient's comprehension abilities enabled her to complete Pierre Marie's Three Paper Test[4] and to correctly mark pictures with oral or written words. The subsequent onset of a right-hemisphere lesion rendered the patient mute and incapable of understanding anything. The anatomopathological examination revealed a left sylvian infarct affecting the deep structure and the parainsular region, as well as another infarct in the territory of the anterior cerebral artery affecting the posterior portion of F1 and the paracentral lobule. The more recent right-hemisphere infarct affected the deep structures of the sylvian region, the parainsular region, and the first temporal convolution. The importance of the initial left-hemisphere lesions and the semiological evolution following the lesions of the right hemisphere suggest an involvement of the right hemisphere in the patient's marked improvement during the first two years following the initial left-aphasiogenic lesions.

Lee, Nakada, Deal, Lin, and Kwee (1984) described the case of a 54-year-old man with Wernicke's aphasia. His aphasia progressively evolved into a moderate amnesic aphasia until the onset, two years later, of a Wernicke's aphasia subsequent to a right-hemisphere infarct of the sylvian region, including the posterior temporal portion and the angular gyrus. The initial lesion had involved the left posterior temporal region and the angular gyrus. It is interesting to note that the second Wernicke's aphasia had a favorable evolution over the next three months. This observation contradicts the findings of Nielsen (1944) and Henschen

---

[4]After handing a small, medium, and large piece of paper to the patient, the examiner asks the patient to throw the large paper on the floor, to put the small paper in his pocket, and to give the examiner the medium one. The patient must listen to the complete command and only then must he execute it in full (Moutier, 1908).

(1926), who, for the sake of functional symmetry, concluded that the contribution of the right hemisphere involved only the areas that were homologous to those of the left hemisphere.

Although the evidence is somewhat indirect, these observations are sufficient to suggest a possible contribution of the right hemisphere to the verbal communication abilities of these patients. However, these observations remain to be confronted with converse cases—which are relatively common but rarely, if ever, published—in which right-hemisphere lesions do not modify the semiology of the aphasia.

Intracarotid Injection of Sodium Amytal

The injection of sodium amytal into the carotid artery is a technique that was introduced by Wada (1949) to selectively anesthetize one hemisphere on a temporary basis. This technique has been used with certain aphasic subjects to identify the hemisphere responsible for the remaining—or in some cases recovered—linguistic skills.[5] For example, a left-side injection in a normal right-hander results in a right-side hemiplegia and a transitory loss of speech, thereby demonstrating the dominance of the left hemisphere for speech (Wada & Rasmussen, 1960).

Kinsbourne (1971a, 1971b) described three aphasic patients in whom an intracarotid injection of sodium amytal produced a transitory loss of speech. The first patient was a 48-year-old man suffering from motor aphasia subsequent to an ischemia in the left sylvian territory. After six weeks of evolution, his expression was limited to a few words but his comprehension was relatively good. Following a right intracarotid injection of sodium amytal, the patient was unable to vocalize or to carry out any buccofacial movement. Conversely, his oral expression remained unchanged following a left intracarotid injection, albeit certain perseverations were present. The second patient was a 27-year-old man with conduction aphasia. Whereas his language behavior remained unchanged by a left intracarotid injection of sodium amytal, a right-side injection left the patient totally speechless and incapable of moving either his lips or his tongue. This observation eventually led Kinsbourne (1972) to consider the compensatory role of the right hemisphere in conduction aphasia, a viewpoint that was later adopted by Levine and Calvanio (1982). The third patient was a 36-year-old man with global aphasia. A left intracarotid injection had no effect on the patient's performance. However, the patient was not administered a right intracarotid injection.

In two of these three cases, a left-side injection did not change the aphasiological semiology, whereas a right-side injection resulted in a transitory loss of speech, thereby suggesting that oral expression is dependent on the right hemisphere. Kinsbourne focused his attention on speech production only. According to Kinsbourne's dynamic model, the left hemisphere is in direct competition with

---

[5]Given that this technique is relatively invasive and not without a certain amount of risk, its application to aphasic populations would probably not be authorized by contemporary ethical committees.

the right hemisphere for the control of speech-related behavior. Consequently, the presence of a left-hemisphere lesion allows the right hemisphere to exercise its limited control over "vocalization" (Kinsbourne, 1971b).

Czopf (1979) distinguished three groups of right-handed aphasic patients based on the effects of an intracarotid injection of barbiturates. In the first group of patients, the injection led to a total suppression of verbal expression. Two of these patients were also administered left-side injections that produced less change in their behavior. The second group consisted of five patients whose aphasias were less severe and more recent than those of the patients in the first group. A right-side injection produced a mild worsening of their aphasias. Four of the patients were also given a left-side injection, which produced similar results. Finally, the third group consisted of three patients with mild aphasia of recent onset (i.e., 2, 10, and 14 days). The right-side injection had no effect on their performance, whereas the left-side injection resulted in a severe aphasia. According to Czopf, these results suggest that the contribution of the right hemisphere is correlated with the amount of time postonset. It should be noted that the deterioration of language behavior following right anesthesia was observed only in the group of patients with the longest time postonset, and also with the most severe aphasia. The role of the lesion size should not be entirely overlooked. The results of this study, like those of Kinsbourne (1971a, 1971b), provide further evidence in support of the right hemisphere's involvement in the residual language of certain aphasic patients.

Although these rare anatomoclinical observations attest to the active role of the right hemisphere in certain aphasic patients, they provide no information about the parameters that determine this participation or about the qualitative aspects of this participation. For example, it is known that several different factors are involved in the recovery from aphasia. These include the size and nature of the lesion, the type of aphasia and its severity, the age of the subject, familial history of left-handedness or ambidexterity, cultural background, knowledge of a second language, and exposure to rehabilitation (Chapey, 1981; Jeannerod & Hécaen, 1979; Seron, 1979; Shewan & Kertesz, 1984; Thiery, Dietens, & Vandereecken, 1982). What is not known, however, is the relationship between these factors and the possible contribution of the right hemisphere.

Given the limitations of sodium amytal studies, and given the importance of these questions, the dichotic listening technique has been considered by several researchers as one way of quantifying the contribution of the right hemisphere to the aphasic's language behavior, in spite of the methodological issues it has raised even in studies with normal subjects (Bradshaw et al., 1986). As we will see, these studies are problematic in many respects.

Dichotic Listening

The administration of a verbal dichotic listening test usually reveals a right-ear advantage in normal right handed adults (Bradshaw & Nettleton, 1983). However, the use of a similar task usually reveals a left ear advantage in aphasic subjects. This finding has been interpreted in two radically different ways (Lambert, 1982a).

Certain researchers (e.g., Moore & Weidner, 1975; Pettit & Noll, 1979) have interpreted the dichotic listening performance of aphasics in much the same way as that of normal subjects. For example, according to this "classical view," a left ear advantage reflects the dominance of the contralateral hemisphere (i.e., the right hemisphere). According to these authors, a shift over time in the dichotic listening ear advantage could be considered as indicative of a more or less complete transfer of dominance for language to the right hemisphere. Other investigators (Schulhoff & Goodglass, 1969) consider the same results as the consequence of a lesion of the left primary or associative auditory cortex, which disturbs the normally prevalent contralateral (right ear) pathway (Sparks, Goodglass, & Nickel, 1970). From this point of view, dichotic listening performance reflects the presence of a lesion but does not allow for any inferences concerning an eventual role of the right hemisphere for language.

To determine the possible contribution of the right hemisphere to the recovery from aphasia, several researchers attempted to correlate dichotic listening performance with the evolution of aphasia over time. These studies have been either *cross-sectional*—comparisons are made between groups of subjects that differ with respect to the time elapsed since the onset of aphasia—or *longitudinal*—the same subjects are compared at different times during the evolution of their aphasia. The longitudinal method allows one to differentiate between group results and individual results, which, in our opinion, is essential in determining if, for a given aphasic subject, there is any evidence of a contribution of the right hemisphere.

Johnson, Sommers, and Weidner (1977) compared the performance of two groups ($N = 10$) of aphasic subjects with left posterior lesions on a consonant–vowel–consonant (CVC) dichotic listening task. The group tested after a delay of six months demonstrated a greater left ear advantage than the group examined less than six months after the lesion onset. Since the subjects with the longest postonset recovery period demonstrated the strongest ear advantage, the authors concluded that this change in dichotic listening performance reflected a contribution of the right hemisphere.

In one of several experiments, Castro-Caldas and Botelho (1980) tested 40 aphasic patients on two separate occasions, once during the first 6 months following the onset of the lesion and once during the next 6 to 12 months. The results differed depending on whether the aphasics were classified as fluent or nonfluent. The tendency for the left ear advantage to increase with time was found only in nonfluent aphasics, and only for words but not for digits. Conversely, fluent aphasics demonstrated a right ear advantage for words. This last result is at odds with those of Johnson et al. (1977), but the stimuli were not the same. According to Castro-Caldas and Botelho (1980), the transfer of dominance to the right hemisphere might occur only in patients with prerolandic lesions. This hypothesis still remains to be verified since the pre- versus postrolandic opposition was based only on fluency criteria and no information was provided concerning the site of the lesions.

Pettit and Noll (1979) administered a dichotic listening task consisting of digits and words to 25 aphasic subjects, twice over a two-month period. The results of the second administration, when compared with those of the first administration two months earlier, revealed a significant increase in the left ear scores, whereas the right ear scores were unchanged. Since the results of an aphasiological assessment indicated signs of recovery, the authors related these two findings and concluded that "there is a strong evidence for the argument that a shift in dominance occurs as the aphasic improves in language ability" (p. 197). This conclusion seems somewhat premature insofar as the authors did not provide any statistical evidence to support a relationship between improvement in language ability and the evolution of dichotic listening performance.

Three recent studies by Niccum and his co-workers (Niccum, 1986; Niccum, Selnes, Speaks, Risse, & Rubens, 1986; Niccum, Speaks, et al., 1986) have provided information lacking in the studies mentioned above. In one study, the dichotic listening performance and the language performance of a group of 54 right-handed aphasics (subsequent to a left ischemic lesion) were monitored during a single semester. The analysis of the individual performance of the 27 subjects who were administered a dichotic listening task on a monthly basis revealed that for the majority of them ($N = 20$), the evolution of their scores did not provide any evidence for a shift in hemispheric dominance (Niccum, 1986). Only in four of the subjects did the evolution of the scores reveal a significant advantage of the left ear over the right, thereby suggesting a transfer of dominance. The evolution of three of these four subjects was compatible with a shift in dominance from the left to the right hemisphere during the course of recovery.[6] Of the 27 subjects, 25 were also evaluated 12 months postonset; their scores remained similar to those obtained 6 months beforehand.

In another study, Niccum, Selnes, et al. (1986) examined the relationship between dichotic listening performance and lesion localization. A significant correlation was found between right ear dichotic listening scores and the presence or absence of a lesion in Heschl's gyrus or in the neighboring cortex. Furthermore, this correlation was higher for scores obtained after six months of evolution. When the identification rate of right ear stimuli exceeded 61%, the probability of Heschl's gyrus being intact was 96%. This correlation suggests that right ear–left ear differences could essentially be dependent on the integrity of the left auditory cortex.

In a third study, Niccum, Speaks, et al. (1986) examined the relationship among dichotic listening performance, aphasic semiology, and lesion localization. Several parameters were taken into consideration, including some from the Boston Diagnostic Aphasia Examination (Goodglass & Kaplan, 1972). The results revealed a correlation between the severity of aphasia and the left ear advantage. However, this correlation was based on very weak right ear scores. Thus the correlation between the severity of aphasia and the left ear advantage, which could be considered as a strong argument in support of a transfer of dominance, could, in fact, be the direct consequence of the left-side lesion. Furthermore, the authors found no correlation between the initial severity of aphasia and the evolution of dichotic listening scores. The quality of individual recovery

[6]One of these cases was the object of a more detailed study by Niccum and Rubens (1983).

was not correlated with the importance of the between-ear difference in dichotic listening performance. Finally, contrary to Castro-Caldas and Botelho's hypothesis (1980), Niccum, Speaks, et al. (1986) found no correlation between dichotic listening performance and the fluency criterion or the pre- versus postrolandic localization of the lesions.

In summary, the results of these studies refute the findings of previous dichotic listening studies suggesting a contribution of the right hemisphere to the residual language of aphasics: "... we believe that dichotic ear advantages do not reflect hemispheric dominance for language processing in brain damaged listeners" (Niccum, Speaks, et al., 1986, p. 314). The author's argumentation is solid enough to suggest that dichotic listening scores reflect, above all, the effects of lesions to the auditory cortex. The results of the last study also suggest that previous studies in which dichotic listening performance was interpreted as indicative of a transfer of dominance to the right hemisphere must be considered cautiously. The fact remains, though, that if one takes into account the limitations of the dichotic listening technique (Bradshaw et al., 1986), as well as the methodological precautions suggested by Niccum and colleagues, then the performance of individual subjects could be used as an argument in favor of a possible contribution of the right hemisphere.

Although the dichotic listening technique does not appear to provide any straightforward answers to the questions raised here, it is possible that additional evidence might emerge from newer functional techniques. Studies using evoked potential recordings (Brandeis & Lehmann, 1986), measures of regional cerebral blood flow (Demeurisse & Capon, 1985; Demeurisse, Verhas, & Capon, 1984), and positron emission tomography (Heiss, Herholz, Pawlik, Wagner, & Weinhard, 1986) have opened the way toward a certain number of solutions. However, because these techniques are currently the target of methodological criticism, it is premature to draw any definite conclusions from them.

Conclusion

The observations presented in this chapter demonstrating that a right-hemisphere lesion can affect recovery from aphasia are rarely found in the literature. In spite of this, they could lead to the belief that the right hemisphere has some potential for language, which can be actualized under certain conditions. However, this potential has been observed only in particular cases and does not necessarily imply that the right hemisphere actually contributes to the recovery of language behavior in all aphasics. This chapter has emphasized the fact that there are few, if any, adequate means of demonstrating the importance and the exact nature of the right hemisphere's contribution to recovery from aphasia in a right-handed adult. However, if these techniques provide results that converge toward the same direction as those from other approaches (e.g., right-brain–damaged studies, divided visual field studies in normals), then these findings could be interpreted in support of the possible contribution of the right hemisphere to recovery from aphasia.

4
The Contribution of the Right Hemisphere to Lexical Semantics

Traditionally, the comprehension and production of words in right-handed individuals have been assumed to be entirely dependent on left-hemisphere processes. However, since the late 1950s a wealth of information has emerged suggesting a possible contribution of the right-hander's right hemisphere to the semantic processing of words. This relatively recent evidence has necessitated reconsideration of the exclusiveness of the left hemisphere's contribution to lexicosemantic processing. The main objective of this chapter is to summarize the evidence from three complementary approaches: studies of split-brain patients, divided visual-field experiments with normal subjects, and studies of patients with acquired lesions of the right hemisphere. Each of these examines, in its own way, the question of the nondominant right hemisphere's contribution to lexical semantics.

Split-Brain Data

Split-brain studies of the right hemisphere's potential for lexicosemantic processing have certainly given an impetus to research in this domain. The first issue examined in this section concerns the capacity of the isolated right hemisphere to process spoken and written words. Next, a series of studies concerning the interconceptual semantic capacities of the right hemisphere is reviewed. Recent investigations of the right hemisphere's sensitivity to semantic priming also are presented. Finally, the evidence from split-brain studies is discussed in terms of the left hemisphere's superiority for lexicosemantic processing as well as in terms of the interhemispheric interactions required for normal lexicosemantic processing.

Comprehension of Written and Spoken Words

The right hemisphere of certain split-brain patients is capable of understanding written and spoken words. It can associate an object with its corresponding name and vice-versa (Gazzaniga, Ledoux, & Wilson, 1977; Sperry, 1968; Sperry &

Gazzaniga, 1967). Thus when the name of an object (e.g., spoon, cup) is presented to the left visual field, split-brain patients often are able to retrieve with their left hand the designated item from among an array of hidden objects. As proof that such behavior depends solely on the right hemisphere, it has been reported that the same subjects are unable to retrieve the designated item with their right hand (Sperry, 1968; Sperry & Gazzaniga, 1967). The right hemisphere is also capable of understanding relatively complex, spoken lexical material (e.g., "used to tell time" for a clock).

Some of the early claims concerning the right hemisphere's capacity for the processing of spoken words must, however, be viewed with caution. Indeed, the stimulus words were presented binaurally, in which case performance could have resulted from left-hemisphere processing due to the involvement of ipsilateral sensory pathways (Gazzaniga, 1967). One way of coping with such problems would be to ensure that the presentation of the visual stimuli and/or the visual array of the possible responses is restricted to the right hemisphere. Furthermore, the use of pointing, instead of manipulating, as a means of responding should be encouraged, thereby minimizing any possible effects of ipsilateral tactile information.

These factors have been taken into account in split-brain studies involving the administration of the Peabody Picture Vocabulary Test (PPVT). For example, this test was used to assess auditory and visual comprehension in patient J.W., whose scores were as good as those of an 18-year-old subject, falling within the upper range of the test norms (Gazzaniga, Smylie, & Baynes, 1984). The auditory comprehension scores of patients N.G. and L.B. were similar to those of an 11-year-old and a 16-year-old, respectively (Zaidel, 1976, 1978a, 1978c). Furthermore, the visual comprehension scores of N.G. and L.B. indicated that the right hemisphere is indeed capable of processing words semantically, as it contributed to a considerable number of semantic errors (Zaidel, 1982). Despite the right hemisphere's lexicosemantic capacities, the left hemisphere's lexicosemantic capacities are far superior (Gazzaniga et al., 1984; Zaidel, 1976, 1978a, 1978c). Moreover, it appears that the right hemisphere has access to a much smaller lexicon than the left hemisphere (Zaidel, 1976, 1978a, 1978c).

The disconnected right hemisphere's capacity to understand spoken words conforms to the word frequency rule,[1] as does the left hemisphere, as well as children, adults, and even aphasics (Zaidel, 1978a, 1978c). Assuming that the frequency effect is somewhat related to the actual linguistic experience of an individual, the right hemisphere's compliance with the word frequency rule suggests that the two hemispheres acquire an auditory lexicon by following similar word-exposure principles (Zaidel, 1978c). However, given the quantitative difference between the words accessible by the right hemisphere and those accessible by the left hemisphere, this acquisition or the capacity to access

[1]According to the word frequency rule, high-frequency words are easier to process than low-frequency words.

this auditory lexicon appears to be somewhat different in each of the two hemispheres. This interhemispheric difference could be attributed to differences in the age of onset of linguistic acquisition processes, as dictated by the structural differences between the two hemispheres (Zaidel, 1978c).

According to Zaidel (1978c), the right hemisphere's auditory comprehension is superior to its written comprehension. In addition, any word that can be read can also be understood in its oral form, though the contrary is not always true. However, Gazzaniga et al. (1984) found no significant differences between the auditory and the written lexicons of the right hemisphere. At least two explanations can account for these contradictory findings. On the one hand, these two studies involved different groups of split-brain patients, and there are important interindividual differences within the general population of split-brain patients (Myers, 1984; Zaidel, 1982). On the other hand, these two studies differed with respect to the exact nature of the tasks used. Whereas Gazzaniga et al. (1984) indicated that their subjects were administered the visual version of the PPVT prior to the auditory version, Zaidel (1978c) neglected to provide such information. Thus one cannot overlook the possibility that the order in which the tasks were presented may have had an effect on the subjects' performance. Despite this eventuality, Zaidel (1978c) suggested that the visual lexicon of the right hemisphere was but a subset of its auditory lexicon.

Zaidel went on to suggest that the right hemisphere does not rely on grapheme–phoneme conversion rules in order to read words. If indeed this were the case, then all spoken words that were understood could also be read. According to Zaidel, the right hemisphere acquires its auditory and written lexicons in different cognitive contexts. Again if this were the case, the psychological and linguistic structure of written words would be slightly different from that of spoken words, although the difference might be very subtle. It is also possible that the right hemisphere's acquisition of a visual lexicon is based on its preexisting auditory lexicon (Zaidel, 1978c). In the absence of grapheme–phoneme conversion rules, the right hemisphere seems capable of understanding written words in terms of specific visual gestalts (i.e., idiographically), which are associated with a particular semantic representation. However, the right hemisphere is capable not only of simple template matching between a given word form and its meaning but also of understanding the same words presented somewhat differently (e.g., by using uppercase or lowercase letters) (Zaidel, 1978c, 1982).

An ideal piece of evidence to support the existence of distinct lexical structures within a given sensory modality is the identification and experimental use of words present in the visual lexicon but not in the auditory lexicon. Zaidel (1978c) reported that he failed to identify such words. However, in a more recent study, Zaidel (1982) observed different patterns of comprehension difficulties as a function of the modality of the stimulus presentation. For example, in one subtest of the Boston Aphasiological Battery, the reading performance of the right hemisphere was best for names of geometric shapes, followed by color names, numbers, actions, and objects in that order. As for the right hemisphere's auditory comprehension, object names were the best understood, followed by names of geometric shapes, actions, color names, numbers, and letters. On the basis of these results, one is tempted to conclude that the right hemisphere does indeed possess differently organized visual and auditory lexicons (Zaidel, 1982). It is also interesting to note that all but one of the right hemisphere's reading errors occurred within the semantic boundaries of the various word categories mentioned above.

One of the more interesting findings to emerge from some of the early studies concerning the disconnected right hemisphere of split-brain patients was that verbs were not represented in this hemisphere (Gazzaniga, 1970, 1971). This result has been disputed by more recent studies, in which it has been shown that when word frequency and the age of acquisition are taken into account, the right hemisphere can understand verbs just as well as nouns (Zaidel, 1978a). It has also been shown that the right hemisphere is capable of understanding visually presented verbs (Sidtis et al., 1981; Zaidel, 1982). Just like the left hemisphere, the right hemisphere can also understand visually presented nouns, choosing the appropriate picture depicting a written action name and executing complex verbal commands such as "Stand like a boxer!" (Gazzaniga et al., 1977). However, although patients P.S., V.P., N.G., L.B., and J.W. demonstrated the ability to comprehend action verbs, when these were presented to the right hemisphere, only patients P.S. and V.P. were able to understand and carry out simple verbal commands (Gazzaniga, 1983a). In light of the right hemisphere's difficulty in carrying out verbal commands (Gazzaniga & Hillyard, 1971; Sidtis et al., 1981; Volpe, Sidtis, Holtzman, Wilson, & Gazzaniga, 1982), it has been concluded that it is more difficult for the right hemisphere to understand verbs than to understand nouns (Sidtis & Gazzaniga, 1983).

There are several hypotheses why verbs and nouns are processed differently. First, noun and verb frequency are seldomly taken into consideration, a methodological flaw that might explain the advantage of nouns (Zaidel, 1978a). Second, certain split-brain patients displayed no initial competence for verb comprehension but did so after a period of time. Thus there could be an influence of the time elapsed since surgery, reflecting a change in the right hemisphere's linguistic ability. Third, tasks involving the comprehension of action verbs may be very different from tasks requiring the execution of verbal commands. Typically, in the former task the subject must choose the appropriate picture illustrating a target word, whereas in the latter task the subject must perform the appropriate gesture. In fact, Levy (see Zaidel, 1985) observed that the disconnected right hemisphere is able to understand verbal commands (e.g., choosing the appropriate picture from a multiple-choice array of pictures), though it is unable to perform the same verbal commands. This did not appear to be a praxic disorder, since the gestures that could not be executed on verbal command could be performed when the subject was presented with an illustration depicting the gesture (Gazzaniga, 1970). The difference between the comprehension of verbs and that of verbal commands containing verbs appears to be limited to the verbal-induced retrieval of gestures. Nonetheless, the right hemisphere does seem to have at least some access to the meaning of a certain number of verbal commands, since it is capable of understanding the instructions of the experimental tasks to which it is submitted (Zaidel, 1985). In this case, however, contextual cues could in part account for the subject understanding.

Another way of examining the disconnected right hemisphere's ability to cope with words is to compare its performance with the performance of aphasic subjects. For example, the mean raw scores of all the disconnected right hemispheres tested on the PPVT have been found to be lower than the mean raw scores of aphasics. However, when the right hemisphere scores were compared with those of various subgroups of aphasics, then the performance of the right hemisphere was found to be more similar to the performance of Wernicke's aphasics than to the performance of the other subgroups (Zaidel, 1976, 1978a). The Word Discrimination subtest of the Boston Diagnostic Aphasia Examination, or BDAE (Goodglass & Kaplan, 1972), has also been administered to split-brain patients as well as to various subgroups of aphasics (Zaidel, 1976, 1978a). In this case, however, the performance of the disconnected right hemisphere bore a closer resemblance to the pattern of scores observed in anomic or Broca's aphasics (Zaidel, 1976). When the right-hemisphere scores were broken down by word category, object names were found to be understood best, followed by geometric shapes, actions, color names, numbers, and letters (Zaidel, 1976). The lower scores obtained for action names than for object names is probably due to the lower mean frequency of words in the former category (Zaidel, 1976). Be that as it may, a similarity exists between the auditory vocabulary of the disconnected right hemisphere and that of aphasics, at least as far as primitive semantic structures are concerned (Zaidel, 1978a).

The performance of the disconnected right hemisphere has also been assessed using the Auditory Comprehension subtests of the Minnesota Test for the Differential Diagnostic of aphasia, or MTDDA (Schuell, 1972). The response cards for these subtests contain fewer pictures to choose from than those of the BDAE. The performance of the disconnected right hemisphere was slightly better than the performance of aphasics but inferior to the performance of the disconnected left hemisphere, which had near-perfect scores (Zaidel, 1976). Here again, the right hemisphere has its biggest problems with isolated-letter items. With regard to "common words," most of the right hemisphere's errors involved confusions between semantically related items (e.g., cup–spoon) or between visually similar pictures (e.g., boy–girl). It is also interesting to note that right-hemisphere errors usually occurred in pairs. For example, if "horse" was confused with "dog," then "dog" was in turn confused with "horse" (Zaidel, 1976).

In a study concerning the reading capacity of the right hemisphere, Zaidel (1982) administered the Western Aphasia Battery, or WAB, to patients N.G. and L.B. as well as to a group of aphasic patients. When the disconnected right-hemisphere scores were compared with those of the various subgroups of aphasics, N.G.'s performance was most similar to that of isolation aphasia, followed by Wernicke's aphasia, and finally Broca's aphasia. For L.B.'s performance, this order was reversed (Zaidel, 1982). Once again, Zaidel (1982) noted a large number of semantic errors (as opposed to visual or auditory errors) by the disconnected right hemisphere. For example, in one of the subtests of the WAB, spoken words had to be matched with written target words. In addition to visual and auditory distractors, the subtest included paradigmatic (e.g., flower–tree, window–door) and syntagmatic (e.g., flower–garden, window–glass) semantic distractors. The results revealed that both N.G.'s and L.B.'s right hemisphere made syntagmatic errors on this particular subtest. According to Zaidel (1982), this suggests that the linguistic contiguity experienced could play an important role with respect to the organization of the right hemisphere's lexicon. Since it has been demonstrated that children produce more syntagmatic responses than paradigmatic responses in word-association tasks, whereas the opposite is true in adults, Zaidel (1982) suggested that the lexicosemantic capacities of the right hemisphere might reflect an earlier stage of language development. Furthermore, the right hemisphere might be more involved in lexical processing in children than in adults (Zaidel, 1982).

In another study, Zaidel (1985) used a word-comprehension task that included a multiple-choice array of response items, all of which were semantically linked in one way or another to the target item (e.g., see Lesser, 1974). Once again, Zaidel observed that the errors made by the disconnected right hemisphere were more syntagmatic than paradigmatic in nature. According to Zaidel, the right hemisphere's lexicon appears to be more connotative than denotative. Finally, using standardized tests, Zaidel (1978c) reported that the right hemisphere is able to understand both concrete and abstract words when they are presented orally. Interestingly enough, this result contradicts the results of studies conducted with normal subjects.

To summarize, the disconnected right hemisphere seems to sustain a certain lexicosemantic organization, or at least it has access to such an organization through the use of subcortical structures (Zaidel, 1976, 1978a). In the latter case, if the right hemisphere does indeed share a lexicon with the left hemisphere, the threshold of its access is somewhat different, namely higher (Zaidel, 1976), or of a different nature.

Interconceptual Semantic Matching

The capacity of the disconnected right hemisphere to match semantically related words, objects, or pictures has been examined in a number of studies. The results indicate that the right hemisphere is sensitive to relations such as cohyponymy (e.g., cherry–apple) (Gazzaniga, 1983a; Gazzaniga & LeDoux, 1978; Zaidel, 1978a, 1978c) as well as to relations of the type "used for" or "part of" (Zaidel, 1978a, 1978c). Moreover, the right hemisphere is capable of making semantic analogies (Zaidel, 1978a). For example, it can recognize from among a set of pictures depicting a bullet, a letter, a pencil, and a picture the item that has the same semantic relation with "envelope" as gun has with holster.

More direct evidence of the right hemisphere's lexicosemantic potential comes from studies of its ability to form interconceptual (semantic) relations in word tasks. In one such study, the sensitivity of N.G.'s right hemisphere to such relations was examined (Sugishita, 1978). The subject was asked to find a word that was semantically related to an object presented in the left hand. N.G.'s right hemisphere was found to be sensitive to four kinds of semantic relations: (1) words that were coordinated to the object (e.g., spoon–fork), (2) words that were contingent to the object (e.g., spoon–cup), (3) words that represented a superordinate of the object (e.g., spoon–silverware), and (4) words that represented an occupation related to the object (e.g., spoon–cook). The only relation that the right hemisphere was not able to form concerned words representing an abstract concept related to the object (e.g., spoon–nutrition). Sugishita offered two explanations for this failure. Either (1) the right hemisphere is unable to extract the meaning of abstract words or to form semantic relations of an abstract nature, or (2) the performance of the right hemisphere is due to subvocal cross-cueing.[2] In the latter case, the right hemisphere is capable of recognizing the object held in the left hand, but only the left hemisphere is capable of extracting the exact meaning of the word presented; at this point, the left hemisphere could inform the right hemisphere in one way or another (e.g., cross-cueing). For some reason, however, cross-cueing does not seem to be as efficient in the case of abstract words.

Semantic matching is not exclusive to N.G.'s right hemisphere. For example, when a word that was semantically related to a target word had to be identified from a set of three words, the right hemisphere of patient P.S. was able to match frequently associated words (e.g., clock–time, porch–house, devil–hell) as well as words with opposite meanings (e.g., army–marine, boy–girl, angel–devil) (Gazzaniga et al., 1977). Patients V.P. and J.W. were administered a task in which a target word was presented in the left visual half-field, followed by a series of four words in free-field presentation. The subjects were instructed to point with the left hand to the word that illustrated one of the following semantic relations with the initial word: (1) synonymy (e.g., boat–ship), (2) antonymy (e.g., day–night), (3) hyponymy (e.g., lake–water), (4) hyperonymy (e.g., tree–oak), and (5) function (e.g., clock–time) (Sidtis & Gazzaniga, 1983). The semantic capacities of the right hemisphere were found to be usually inferior to those of the left hemisphere. Moreover, the performance of J.W.'s right hemisphere was always inferior to that of V.P.'s, whereas the performance of the two left hemispheres was almost identical. The performance of both J.W. and V.P. varied very little from one semantic relation to another. For four of the five relation types, the intertype variation was 4 percent for patient V.P. and 10 percent for patient J.W. (Sidtis & Gazzaniga, 1983). Antonymy was the most difficult relation for V.P., whereas synonymy was the most difficult one for J.W. Not only were these semantic relations associated with the poorest right-hemisphere performance, but they were also the ones for which the left hemisphere was the least competent. Thus this difference between the right and left hemisphere appeared to be more quantitative than qualitative.

---

[2]This phenomenon refers to the strategies used by the subject so that one hemisphere can cue the other by way of subvocal messages or behaviors.

In another study, Gazzaniga et al. (1984) administered three language tasks to patients J.W. and V.P. These tasks were designed to examine the semantic fields of four target words: "ant," "eye," "bed," and "bunny." In the first task, the target word was presented in either the left or right visual field. The subjects were asked to choose which of two test words, presented in either the same visual field or the contralateral field, was more closely associated with the previously presented target word. In the second task, the subjects had to indicate if there was an association between the target word, presented in either the right or left visual field, and a test word that was subsequently presented in either the same or the opposite visual half-field. The third task was somewhat similar to the second, except that the two words were presented simultaneously: the target word appeared at the point of fixation while the test word appeared in either the right or the left visual half-field. Once again, the subjects had to indicate if the test word was associated with the target word. The results of this study revealed that the subjects were performing at chance level in the contralateral conditions of the first two tasks, that is, when the target word and the test word were presented in opposite visual half-fields (Gazzaniga et al., 1984). One can therefore conclude that there was no efficient interhemispheric transfer in these two conditions. Contrasting with these results was the finding that when both the target and the test words were presented in the same visual half-field, the right hemisphere seemed to be sensitive to the following semantic relations: hyponymy (e.g., ant–insect), cohyponymy (e.g., ant–bee), attribute (e.g., ant–small), and function (e.g., ant–crawl).

In fact, V.P.'s right hemisphere performed almost as well as his left hemisphere in the first task. The performance of J.W.'s right hemisphere was inferior to the performance of his left hemisphere for the first two tasks only. Performance of the right hemisphere was better for the third task than for the second task, and better for the second and third tasks than for the first task where performance was at chance level. On the basis of these results, it can be said that the degree of the right hemisphere's semantic competence varies from one subject to another. In this case, V.P.'s right-hemisphere lexicon appears to be more efficient and/or complete than J.W.'s lexicon. However, in both subjects, the right-hemisphere lexicon appears to be smaller and not as rich as the left-hemisphere lexicon (Gazzaniga et al., 1984). It might be that the lexicons of the two hemispheres are similar but that access to the right-hemisphere lexicon is more difficult. It could also be that this apparent asymmetry is in fact a reflection of the particular nature of the semantic judgments required in these tasks (Gazzaniga et al., 1984). The results for task three clearly indicate that the nature of the task had an effect on the semantic judgments of J.W.'s right hemisphere (Gazzaniga et al., 1984).

Zaidel (1982) administered the Sklar Aphasia Scale to patients L.B. and N.G. in order to examine the right hemisphere's capacity to find the antonym of a written target word. The target antonym was presented visually along with a set of distractors that were either synonyms or semantic associates of the target word (e.g., hot: warm, cold, sun). Zaidel (1982) reported that the two right hemispheres performed well above chance level and that the performance of L.B.'s right hemisphere was perfect. Interestingly, three of the four errors made by

N.G. involved syntagmatic associations (e.g., white → light) rather than paradigmatic associations (e.g., white → gray). These errors should be added to the verbal comprehension errors described in the previous section. All in all, these findings should encourage further hypotheses concerning the right hemisphere's lexicosemantic organization or strategies.

For the time being, as far as verbal comprehension is concerned, it appears that the limiting factor of the disconnected right hemisphere is not linked to the nature of the semantic representation that can be accessed by this hemisphere (Sidtis & Gazzaniga, 1983; Sidtis et al., 1981). The poor performance of the right hemisphere can be explained in a number of ways. For example, it may be that the right-hemisphere lexicon is smaller than the left-hemisphere lexicon, or that the right hemisphere is less efficient in processing visually presented words (visual recognition), or that it is more difficult for the right hemisphere to respond to verbal material (Sidtis et al., 1981). According to Sidtis et al. (1981), the first of these hypotheses—that of a more restricted lexicon—is unlikely since a word-induced semantic activation in one "disconnected" hemisphere facilitates the processing of a semantically related word in the other hemisphere. These investigators argue that both hemispheres have access to a functionally common semantic network.

Sensitivity to Semantic Priming

The semantic priming effect refers to the influence that the perception of a first element (the prime) has on the processing of a second element (the target) in instances where a semantic association exists between the two elements. For example, in a lexical decision task, the prior presentation of the word "cat" will reduce the amount of time necessary to recognize that a series of letters such as d–o–g forms a word of the English language. When the stimuli are presented laterally, semantic priming can provide an evaluation of the semantic capacities of each hemisphere.

In one semantic-priming task (Sidtis & Gazzaniga, 1983), J.W. had to judge whether a target word designated something natural or artificial. The primes and the targets were presented in either the same or the opposite visual half-field. An intrahemifield effect and interhemifield effect were observed for the targets presented to the right hemisphere. Furthermore, the intrahemifield priming effect did not differ significantly from the interhemifield priming effect. The presence of an interhemifield priming effect suggests that subcortical structures may play an important role in the interhemispheric transfer of information (Gazzaniga, 1983a). This subcortical transfer concerns only semantic representations since J.W.'s right hemisphere has no access to the phonological representation of words (Gazzaniga, 1983a). When the corpus callosum is completely sectioned, each hemisphere maintains access to a functionally common semantic system (Sidtis & Gazzaniga, 1983). The linguistic systems that are accessible by each hemisphere are therefore not entirely independent functional units. The limitations of the right hemisphere with respect to word comprehension are related to the processes involved in the access to this semantic system. It remains to be explained why, in spite of an interhemispheric interaction, neither of J.W.'s hemispheres is able to name stimuli presented to the right hemisphre. Sidtis and Gazzaniga (1983) have suggested that the semantic activation initiated in one hemisphere might facilitate the decision making of the other. However, this process does not provide the right hemisphere with enough information to correctly name the stimulus.

Zaidel (1983a, 1986) investigated the effects of semantic priming within the context of a lexical decision task (word–nonword). Once again, the semantic primes and the target words were presented either in the same or the opposite visual half-field. Four of the six subjects who took part in the study had undergone complete commissurotomies; the other two subjects presented with partial commissurotomies. L.B. was the only subject whose performance had significantly benefited from the semantic priming. Moreover, this priming effect was observed only for L.B.'s left hemisphere, in spite of the fact that other commissurotomized patients are capable of making lexical decisions using either one of their hemispheres. Thus when targets and primes are presented visually, the disconnected right hemisphere's sensitivity to semantic priming is inferior to that of the normal right hemisphere, even though these two hemispheres exhibit similar competence with respect to lexical decision (Zaidel, 1986a). An oral version of the lexical decision task with semantic priming has also been used. Whereas the auditory prime is now perceived by both hemispheres, the visual target continues to be perceived by one hemisphere or the other (Zaidel, 1983a). Semantic priming has a similar effect on both hemispheres: four out of five right hemispheres and four out of five left hemispheres. Taken together, these results suggest that the right hemisphere is sensitive to semantic priming when the prime is auditory but not when it is visual (Zaidel, 1985). Once again, the visual vocabulary of the right hemisphere appears to be but a subdivision of this hemisphere's auditory lexicon (Zaidel, 1985).

In summary, there is some controversy surrounding the sensitivity of the right hemisphere to semantic priming when the stimuli are visually presented. Some of the factors that could account for the divergent findings include the subjects themselves (e.g., J.W. was the only subject in the Sidtis and Gazzaniga study but did not participate in the Zaidel experiments), the nature of the tasks (semantic decision versus lexical decision), and the possibility of residual interhemispheric transfer in certain subjects. However, despite this controversy, it remains clear that the disconnected right hemisphere appears sensitive to semantic activation induced by orally presented words.

Other Manifestations of Lexicosemantic Competence

Gazzaniga and Smylie (1984) assessed V.P.'s and J.W.'s knowledge of current events. The purpose of the study was to examine each hemisphere's acquisition of new information (i.e., current events) following commissurotomy. A word was presented to either the left or the right visual half-field and the subjects had to select an associated word from a set of six written choices. For example, following the presentation of the word "hostage," the subjects had to decide which one of the following words was associated with the target word: Iran, Russia, England, Australia, Cuba. The association to be made here concerned the taking of American hostages in Iran. It is interesting to note the connotative nature of the meaning given to the target word; in other words, the word "hostage" is not associated with Iran a priori. Both V.P.'s and J.W.'s right hemisphere and left hemisphere performed equally well on this task. Moreover, the two hemispheres tended to miss the same trials. This similarity in the performance of the two hemispheres is amazing. Unfortunately, Gazzaniga and Smylie (1984) neglected to state whether cross-cueing strategies were not used.

Hemispheric Dominance: Competence, Superiority, and Control

The concept of hemispheric dominance can refer to the exclusive nature of one hemisphere's contribution, to the superiority of one hemisphere's performance over the other, or to the control that one hemisphere has on the global behavior of an individual when the two hemispheres interact (Zaidel, 1978b). The results from studies of commissurotomized subjects suggest that the lexicosemantic competence of the left hemisphere is superior to that of the right, though it is not entirely exclusive to the former. The control of semantic processing of words has yet to be clearly established. However, a study conducted by Levy and Trevarthen (1977) provided some preliminary findings concerning this issue.

Levy and Trevarthen (1977) used a series of four-letter words (e.g., lady, ball, shoe) to create a set of chimeras by joining the first two letters of one word with the last two letters of another (e.g., lall, baoe). The first two letters and the last two letters of each chimera were presented in the opposite visual hemifield. The subjects were required to point to the picture representing the word they had seen. If, for example, the stimulus word is "baoe," the subject might point with his right hand to the picture of a shoe if performance on this task is controlled by the left hemisphere. If the right hemisphere controls performance on this task, then the subject might point with the left hand to the picture of a ball. Results showed that, regardless of the hand used, the subjects had a preference for information presented in the right visual hemifield. The left hemisphere seems to assume control over semantic processing of words when the two hemispheres are stimulated simultaneously (Levy & Trevarthen, 1977). However, beyond these findings, it remains to be determined if an intact right hemisphere is required for optimal performance on a task such as this.

Two Hemispheres Are Better Than One

The disconnected left hemisphere often appears exempt of any linguistic deficit and is therefore sufficient for semantic processing of words (Gazzaniga, 1970;

Gazzaniga & Sperry, 1967; Zaidel, 1976). However, there is some evidence suggesting that an interhemispheric contribution is necessary for optimal lexicosemantic processing (Zaidel, 1978a, 1978c). For example, on the receptive subtests of the Illinois Test of Psycholinguistic Abilities (ITPA), patients N.G. and L.B. obtain better scores in free vision than in lateralized vision (Zaidel, 1978a, 1978c). Similar findings have been obtained with Form A of the Peabody Picture Vocabulary Test (Zaidel, 1976, 1978a). Interestingly, the performance of commissurotomized subjects was similar to the performance of hemispherectomized subjects tested in free vision (Zaidel, 1976, 1978a). Thus the observed pattern does not appear to be dependent on a negative bias caused by the lateral presentation of the stimuli, but instead suggests the existence of a residual subcortical interhemispheric interaction in commissurotomized subjects (Zaidel, 1978a, 1978c), as well as of a contribution of the right hemisphere (Zaidel, 1976).

A task used by Levy (1978) involved naming the characteristic color of objects depicted in line drawings. The results revealed that the performance of the disconnected left hemisphere of patients L.B., N.G., and C.C. was inferior to that of a normal subject. This finding suggests that the left hemisphere has difficulty in forming a representation of an object with its usual colors (Levy, 1978). The role of the right hemisphere is therefore essential in order to perform a task such as this. Insofar as the color of an object has a semantic value, it would appear that the right hemisphere does provide the left hemisphere with a certain amount of semantic information.

Conclusion

In summary, studies of ccommissurotomized patients suggest that the activity of the right hemisphere in certain right-handers underlies a number of processes contributing to semantic processing of words. The isolated right hemisphere seems capable of reading and of understanding orally presented words and is also sensitive to the presence of semantic relations between words. However, this lexicosemantic potential of the right hemisphere is usually inferior to that of the disconnected left hemisphere. Furthermore, when the two hemispheres are simultaneously called upon to contribute to a task with lexicosemantic requirements, it is usually the left side of the brain that assumes control of the task. Nonetheless, certain arguments support the notion that two hemispheres are better than one. It is therefore reasonable to assume that the right hemisphere not only possesses a certain lexicosemantic potential, but that it also contributes to lexicosemantic processes. However, if residual interhemispheric transfer were to be definitely identified in split-brain subjects, then this potential and/or contribution of the right hemisphere would certainly be less important than it appears now. In addition, this potential and/or contribution could be exceptional, specific to a certain number of commissurotomized subjects, not necessarily reflecting the contribution of the normal intact right hemisphere. If this were indeed true, then the lexicosemantic competence of the disconnected right hemisphere would have to be considered as the superior limit of this hemisphere's potential under optimal conditions (Zaidel, 1978a).

Studies of Normal Subjects

The capacity of the nondisconnected, nondamaged right hemisphere to recognize and identify as words groups of letters is a controversial issue (Hannequin, Goulet, & Joanette, 1987). Nevertheless, even if one were to admit that this capacity does indeed exist, it remains to be demonstrated that the right hemisphere possesses the necessary competence to attribute meaning to these groups of letters. The evidence supporting such competence comes from observations of the normal right hemisphere's sensitivity to the imageable and/or concrete nature of words. This evidence is based also on the findings from two experimental paradigms: studies of semantic judgments and studies of the influence that one stimulus can have on the processing of a semantically similar stimulus.

The Concrete and Imageable Nature of Words

One of the first parameters examined in divided visual field studies with normal subjects was the abstract–concrete dichotomy. After presenting abstract and concrete words in either visual hemifield, Ellis and Shepherd (1974) made the classic observation that words presented in the right visual field were usually recognized better than words presented in the left visual field, at least when they were orally reported. Furthermore, concrete words were significantly easier to identify than abstract words, so long as they were presented in the left visual field. Thus the functional asymmetry in word processing appears to be most evident for abstract words. In explaining their findings, Ellis and Shepherd hypothesized that certain concrete words are represented bilaterally, either because they are usually acquired early in life or because they possess a high degree of imagery, in which case they are more easily processed by the right hemisphere. Ellis and Shepherd did not believe that interhemispheric transfer was a logical explanation for their findings. Indeed, there seemed to be no reason why the neural representations of concrete words would be less resistant than the neural representations of abstract words to transfer from the right hemisphere to the left hemisphere, unless of course they were recognized beforehand by the right hemisphere.

The importance one should give to the degree of concreteness in words processed by the right hemisphere was also demonstrated by Day (1977, experiment 1) in a lexical decision reaction time task. Results revealed no difference between hemifields for speed of response to concrete words. However, there was a right visual hemifield superiority for speed of response to abstract words. According to Day (1977), lexical entries representing abstract words are accessed more easily in the left hemisphere, whereas lexical entries representing concrete words are accessed as easily in either of the two cerebral hemispheres. Unfortunately, the significance of this interpretation is diminished somewhat by the fact that there was no statistical analysis of the supposed interaction between word concreteness and visual hemifield. However, in a recent replication of Day's study, Mannhaupt (1983, experiment 1) reported a statistically significant word concreteness by visual hemifield interaction. Specifically, there was a significant difference between visual hemifields for abstract nouns but not for concrete nouns.

More recently, Restatter, Dell, McGuire, and Loren (1987) compared imageable and frequent concrete and frequent abstract words in a lexical decision task involving a verbal go–no-go response rather than a nonverbal response. When a series of letters formed a word, subjects were instructed to produce the letter *a*. When a series of letters formed a nonword, subjects were instructed not to say anything. The stimuli were presented in either the left or the right visual hemifield. The analysis of reaction time data revealed a significant word concreteness by visual hemifield interaction. Restatter et al. also reported a double dissociation in that abstract words were processed more rapidly than concrete words in the left visual hemifield, whereas the opposite was true for words presented in the right visual hemifield. The authors concluded that the right hemisphere was capable of processing both concrete and abstract words, and they attributed the double dissociation to interhemispheric inhibitory mechanisms.

The relationship between the concreteness of certain words and the possibility that such words are processed by the right hemisphere has not received unanimous support. For example, Searleman (1983) reported that Orenstein and Meighan (1976) failed in their attempt to replicate the findings of Ellis and Shepherd 1974). Although Orenstein and Meighan did obtain an interaction between the degree of word concreteness and the performance of each cerebral hemisphere, it was not statistically significant (Coltheart, 1980). Saffran, Bogyo, Schwartz, and Marin (1980) also failed to obtain a significant word concreteness by visual hemifield interaction. In this study, subjects had to respond either in writing (experiment 1) or orally (experiment 2). Shanon (1979, experiment 2) compared concrete and abstract nouns using a lexical decision task in which word frequency was carefully controlled. The analysis of reaction time data and error data revealed no significant differences between the two visual hemifields. Furthermore, no interaction was found between word concreteness and visual field of presentation. Insofar as Day's experiment was practically identical to his own, Shanon was able to find only one methodological detail that could explain the difference between the results of the two studies: Day's study was somewhat longer. Thus it was possible that the lateralization effect reported by Day (1977) was due to the differential sensitivity of the two cerebral hemispheres to practice, particularly with respect to the processing of concrete words by the right hemisphere. However, both Shanon and Mannhaupt (1983) were unable to demonstrate that visual field differences for concrete nouns diminished with practice.

Bradshaw and Gates (1978, experiment 3) examined the oral production of words presented in either of the visual hemifields. The results revealed that the right visual hemifield advantage for word processing was greater for abstract words than for concrete words, reaching a maximum for low-frequency abstract words and a minimum for high-frequency concrete words. However, these observations were valid for error data but not for reaction time data. Although the analysis of accurately reported words indicated a certain aptitude of the right hemisphere to process concrete words,[3] the analysis of the reaction time data led Bradshaw and Gates to conclude that there was no evidence of an eventual contribution of the right hemisphere to the procesing of frequent concrete words in right-handers.

In spite of contradictory findings to the effect that the right hemisphere is capable of processing certain words, it would appear that such processing is affected by the degree of word concreteness. However, the results of experiments conducted by Hines (1976, 1977) suggest that a word's concreteness is not a sufficient characteristic for it to be recognized by the right hemisphere.

Hines (1976) considered the degree of word concreteness (i.e., concrete versus abstract) as well as the frequency of word usage (i.e., frequent versus less frequent). In this experiment, three-letter words were presented tachistoscopically (20 ms), either unilaterally or bilaterally in pairs. The subjects were instructed to orally identify a digit presented at the point of fixation and then the word or words they had seen to either side of this digit. The main finding of this study shed further light on the findings of Ellis and Shepherd (1974). Indeed, Hines found a larger right visual hemifield advantage for abstract words occurring only for frequent nouns. As far as less frequent words were concerned, it would appear that the right visual hemifield advantage was unaffected by their degree of concreteness. This suggests that the right hemisphere is capable of processing frequent concrete words but has little or no capability to process less frequent words.

In a second study, Hines (1977) replaced the dichotomous oppositions of his earlier study (i.e., concrete versus abstract, frequent versus less frequent) with three levels of concreteness and frequency. The results again revealed a right visual hemifield advantage. Moreover, for high- and medium-frequency words, the left-hemisphere advantage was greater when the stimuli were abstract than when they were concrete or moderately concrete. According to Hines, this suggests that certain words (i.e., concrete and frequent) presented in the left visual hemifield can indeed be recognized by the right hemisphere.

If the right hemisphere is indeed capable of processing words, it would appear that these words are either concrete or, more specifically, concrete and frequent. In addition to word concreteness and frequency, word imageability should also be taken into account. Paivio, Yuille, and Madigan (1968) defined the degree of imageability associated with a word as the ease with which the word evokes a sensory experience in the form of a mental representation of images, sounds, odors, and the like. In contrast, a word is said to be concrete if it refers to objects that can be directly perceived by the senses; a word is said to be abstract if it refers to a concept that is difficult if not impossible to experience by way of the various sensory receptors (Paivio et al., 1968; Toglia & Batig, 1978). Thus the difference between concreteness and imageability is a function of the mental representation characterizing the imageable nature of a word.

---

[3]According to Bradshaw and Gates (1978), the number of errors was "dangerously small for statistical analysis."

Imageability and concreteness frequently are confounded in studies of the functional lateralization of the brain for language. Often they are treated as if they were quasi-synonymous (e.g., Bradshaw & Gates, 1978; Mannhaupt, 1983; Saffran et al., 1980). Although imageability and concreteness are positively correlated (Paivio et al., 1968; Toglia & Batig, 1978), the two factors are dissociable. The words "anger" and "antitoxin" can be used as examples to illustrate this dissociation (Boles, 1983). An antitoxin is something that is difficult to visualize (i.e., higher degree of concreteness than imageability), whereas anger is not a concrete thing, though it evokes mental imagery more easily (i.e., higher degree of imageability than concreteness). Within the context of the functional lateralization of the brain, imageability and concreteness effects should be considered separately.

Marcel and Patterson (1978) reported that the degree of imageability had an effect on word recognition in the left but not in the right visual hemifield, whereas the degree of concreteness had no effect at all. In this experiment, the subjects had to orally report a word presented in either of the visual hemifields. A visual masking technique was used to make the recognition task more complex. As a result, very few low-imagery words were recognized when they were presented in the left visual hemifield. In a second experiment, Marcel and Patterson again observed an imagery effect even when word frequency, length, and concreteness were taken into account. In explaining their findings, Marcel and Patterson suggested that semantics be divided depending on whether it refers to the sensorimotor domain or is based on logical or linguistic concepts. According to this viewpoint, the imageable aspect of an element is based on the sensory experiences it evokes and is therefore related to the sensorimotor domain of semantics. Ontogenetically, sensorimotor-based semantic representations are integrated before logicolinguistic-based representations. This division interacts with the gradual functional lateralization of the brain. According to this hypothesis, each semantic component is represented within each of the cerebral hemispheres. However, while the sensorimotor-based semantic representations in the right hemisphere have a direct access to the left hemisphere's linguistic production system, the right hemisphere's logicolinguistic component has but an indirect access to this production system (output lexicon) by way of its functional equivalent in the left. Under these conditions, when the quality of the stimulus is poor, as is the case here, words whose semantic component has but an indirect access to the production system are the most affected. In other words, the least imageable words presented to the right hemisphere seem the least recognized when the task requires their oral production. The right hemisphere's "problem" is therefore one of production rather than one of a capacity to process words that are low in imageability.

Boles (1983) conducted a series of experiments using word-recognition tasks in which imageability, concreteness, and familiarity were manipulated. The subjects were instructed to orally report the words, which were briefly presented (75–100 ms) in either the left or right visual hemifield. However, Boles was unable to find any evidence that imageability or concreteness had an effect on either the functional asymmetry of the brain or the overall performance of the subjects. The two major findings of this study were a robust right visual field advantage and a general effect of familiarity on word recognition. Moreover, an interesting step in Boles's analysis of all five experiments was the identification of eight words that were high in concreteness, imageability, and familiarity. The right visual advantage was highly significant for all eight words. Boles concluded that only familiarity had an effect on near-threshold word recognition, whereas concreteness and imageability had an effect on performance only after word recognition had occurred. Furthermore, Boles suggested that although the word recognition performance of the left hemisphere was superior to that of the right hemisphere, the lexicons of the two hemispheres were similar in terms of concreteness and imageability. Be that as it may, it should be noted that the left-hemisphere advantage reported by Boles may have been the result of subjects having to orally produce the recognized words.

Unlike the procedure employed by Marcel and Patterson (1978) and Boles (1983), Day's (1979) second experiment did not require the oral production of the recognized words. The purpose of this study was to answer the following questions: Is the right hemisphere capable of processing adjectives and verbs as well as nouns? If so, is this capacity linked to the imageable nature of these words? Two groups of normal subjects were administered a lexical decision task consisting of nouns, adjectives, and verbs that differed in terms of their degree of mental imagery (i.e., high imagery versus low imagery). Results showed a right visual hemifield advantage for nouns and adjectives having a low degree of imagery and for verbs regardless of their degree of imagery. According to Day, these findings suggest that the lexicosemantic processes sustained by the right hemisphere are dependent on the degree of imagery and the grammatical class of the word stimuli. The right hemisphere is capable of processing highly imageable nouns and adjectives, whereas less imageable nouns as well as verbs, regardless of their degree of imageability, are processed almost exclusively by the left hemisphere. Day suggested that word recognition by the right hemisphere is somewhat facilitated by an internal lexical code based on imagery, and that verbs, even the most imageable ones, are not as easily integrated by this type of code as are nouns and adjectives. To further emphasize the particular status of verbs, Day cites Reid (1974, in Day, 1979), who postulated that verbs are not directly represented by images but instead are indirectly represented by the relations that exist between subjects and objects.

Unlike the conclusions of Marcel and Patterson (1978), Day (1979) suggested that the production problem is not one of access to a production system but depends on the limits of the right hemisphere to process words having a low degree of imageability. However, there is one point that should be emphasized in this study and that is the failure to control the concreteness factor. Indeed, the highly imageable words used by Day were also highly concrete, whereas the less imageable words were also less concrete. It is therefore difficult to dissociate the relative contribution of concreteness and imageability factors in this study.

Lambert and Beaumont (1983) conducted four separate experiments in which they manipulated the degree of word imageability and that of word concreteness, as well as several other variables including the orientation of the stimulus presentation (i.e., horizontal versus vertical) and the order of report (i.e., free report, left first, right first). The word stimuli were presented either unilaterally or bilaterally. The results of the experiments revealed that, contrary to the findings of Marcel and Patterson (1978) and Day (1979), the right visual hemifield advantage was not affected by the degree of word imageability or concreteness. Lambert and Beaumont attributed the results of previously published studies supporting the right-hemisphere imageability hypothesis to various experimental artifacts. Similarly, Schmuller and Goodman (1979) reported that they had found no evidence to suggest that high-imagery words were lateralized any differently than low-imagery words. In their study, the subjects were instructed to report pairs of bilaterally presented words in a particular order. The results (error data) show a larger right visual hemifield advantage for frequent, highly imageable words than for high-frequency words having a low level of imagery. The results also reveal that processing was easier for high-imagery words than for low-imagery words in both visual fields. Schmuller and Goodman's interpretation suggests that the right hemisphere is incapable of word processing.

More recently, Howell and Bryden (1987) suggested that when the frequency of use and the spatial display of word stimuli are controlled, then the functional lateralization of the brain for word processing is not affected by the imagery factor. This conclusion was based on the results of a lexical decision task presented in lateralized vision. To explain the lack of any influence of word imageability, Howell and Bryden referred to Paivio's (1971) dual-coding model. According to this model, words are mediated by two independent though partially interconnected cognitive systems: the verbal system and the imagery system. Within such a context, the degree of imageability associated with a particular word is indicative of the association between the word and its image. Thus it seems normal that subjects' performance be affected by word imageability only after the word has been recognized – in other words, after it has been processed by the left hemisphere.

McMullen and Bryden (1987) sought to provide support for the supposed right hemisphere's potential to process highly imageable words by administering a lexical decision task to female subjects and instructing them to provide a manual response with the left hand. The rationale for using such a procedure is that the functional lateralization of the brain for language is claimed to be less well established in women than in men and that responding with the left hand requires activation of the right hemisphere. The degree of word imageability and the frequency of word usage were varied and the stimuli presented unilaterally and horizontally. Despite their efforts, McMullen and Bryden failed to observe that imageability and word frequency affected the right visual hemifield advantage for word processing. Neither the analysis of accuracy scores nor that of reaction time data provided any evidence suggesting that words which are high in imageability are represented in the right hemisphere. McMullen and Bryden did observe, however, that frequently used words were processed more easily than infrequent words, and that high-imagery words were processed more easily than low-imagery words.

Unlike many of the previous researchers, Bruyer and Strypstein (1985) admitted that imageability and concreteness were relevant to the lateralization of the brain for language. However, they argued that the influence of these two factors varied according to the experimental task to be performed. The evidence for this came from three experiments. In one experiment, the subjects were given a lexical decision task. Both error data and reaction time data revealed the usual right visual hemifield advantage. Moreover, this advantage was not affected by imageability or concreteness when either of these two factors was maintained at a moderate level. In a second experiment, another group of subjects was given a judgment task in which they had to match a spoken word with a word presented in either visual hemifield. Statistical analyses revealed the usual right visual hemifield advantage when the degree of word concreteness was varied and the degree of word imagery maintained at a moderate level. However, this advantage disappeared when the stimulus words were of an average degree of concreteness or of variable imagery. Finally, in a third experiment, the subjects had to indicate if laterally presented words were included in a list of words previously presented in central vision. The results revealed no significant difference between the two visual hemifields when the words were varied in concreteness but maintained constant with respect to their degree of imageability (moderate level). In contrast, a right visual hemifield advantage was obtained when the stimuli were varied in terms of their imageability and maintained constant in terms of their concreteness (moderate level). However, this advantage resulted from the extreme slowness in processing highly imageable words presented in the left visual hemifield.

According to Bruyer and Strypstein (1985), as well as Marcel and Patterson (1978), imageability is the factor most likely to influence or explain the functional lateralization of the brain for word processing. In reality, the results of their experiments suggest that the influence of imageability and concreteness on the left-hemisphere advantage for language processing depends on the nature of the task to be performed. The results of their first two experiments did not demonstrate the visual hemifield by nature of the word interaction that is required if one is to assume that the right hemisphere does indeed possess a certain linguistic competence. The results of their second experiment showed that low-imagery words were processed better when they were presented in the right visual hemifield, though high-imagery words were processed just as quickly when presented in either the left or right visual hemifield. However, this quasi-interaction was not supported by the statistical analyses. Only the results for the third experiment revealed this interaction, and the variable involved then was imageability. Even if one were to agree with Bruyer and Strypstein's conclusion that the right hemisphere is superior to the left in terms of mental imagery processes, the fact remains that this interaction is due to the slower processing of high-imageable words presented in the left visual hemifield. Thus Bruyer and Strypstein's observations and conclusions must be considered cautiously. Furthermore, their stimulus words were described as being "highly saturated in frequency," leading the authors to believe that the frequency factor had been adequately controlled. As Bruyer (personal communication 1988) later acknowledged, there is no guarantee that this factor was actually controlled. Thus attempts should be made to replicate the findings of this study.

Another stimulus factor that has been related to imageability is word length. According to Young and Ellis (1985), several studies have shown that the right visual hemifield advantage is greater for longer words (e.g., Bouma, 1973; Gill & McKever, 1974; Melville, 1957; Schiepers, 1980; Turner & Miller, 1975). None of these studies, however, controlled the degree of imageability of the stimulus words. Thus it is possible that the influence of word length may have stemmed from the fact that the longer words had a lower degree of imageability than the shorter ones.

To examine this hypothesis, Young and Ellis (1985) conducted a series of eight experiments concerned with the recall of bilaterally presented words and pronounceable nonwords. The manipulated factors included not only word length and the degree of word imageability, but also word frequency and the format used to display the letters forming each word. The results revealed that for pronounceable nonwords and words displayed in a nonconventional format (i.e., vertical alignment or words poorly aligned in the horizontal axis), word length influenced recall in both visual hemifields. In contrast, for words displayed in a conventional format (i.e., letters appropriately aligned in the horizontal axis), word length affected performance in the left visual hemifield only. The interaction between word length and visual hemifield was observed regardless of the degree of imageability or the frequency of use of the word stimuli.

Young and Ellis (1985) also found an interaction between word length and the imageability and frequency factors. In fact, as far as left visual hemifield presentations were concerned, performance was better for words than for nonwords, for frequent words than for less frequent words, and for high-imageable words than for low-imageable words. In their discussion and conclusions, Young and Ellis argued that lexical access was somewhat possible by way of the left visual hemifield. One piece of evidence for this comes from the better recall of words than nonwords. Another piece of evidence is the influence of imageability and frequency on the processing of words presented in the left visual hemifield. The differences observed between the two visual hemifields—in particular the interaction between the visual hemifields and word length when words are presented horizontally—attest to the presence of different modes of lexical access. Young and Ellis also admit that there may be a qualitative difference between the two cerebral hemispheres at one or several levels during the processing of linguistic stimuli. However, they did not indicate where the lexicon acccessed by words presented in the left visual hemifield is located. It is possible that the right hemisphere has a lexicon of its own, or that it has access to the left hemisphere's lexicon, or both.

In a more recent study, Bub and Lewine (1988) criticized Young and Ellis (1985) for presenting their stimuli bilaterally and using a forced order of report. Bub and Lewine (experiment 4) reexamined the influence of word length on the functional lateralization of the brain for language in a lexical decision task where the stimuli were presented unilaterally. They manipulated not only word length but also the degree of concreteness, based on the assumption that concreteness reflects the degree of word imageability. The stimulus words were matched for frequency of use. The results showed that for stimuli presented in the right visual hemifield as well as in central vision, processing times did not vary as a function of word length or degree of concreteness (concrete–abstract). Conversely, for stimuli presented in the left visual hemifield, word length influenced processing speed, though much less so in the case of concrete words, for which the effect was not significant. Bub and Lewine concluded that the degree of word imageability had an influence on the processing of words presented in the left visual hemifield. In addition, the speed of decision making for abstract words is strongly influenced by word length but only when words are presented in the left visual hemifield. Thus the classic right visual hemifield advantage increases with the length of stimulus words when they are abstract words but not when they are concrete. According to Bub and Lewine, concreteness influences word proccessing, but this effect depends on word length: in comparison to shorter words, longer words result in a greater difference between the processing of concrete words and that of abstract words. It is probable that the use of shorter words explains why several previous studies failed in their attempt to demonstrate an interaction between concreteness and the visual hemifield of presentation. In summary, Bub and Lewine conclude that highly concrete words can be directly processed by the right hemisphere. However, they add that even though the right hemisphere may have a certain level of lexical access, words may ultimately have to be transferred to the left hemisphere to be processed. It is possible that the only lexical representation sustained by the right hemisphere corresponds to a series of graphic symbols which then have to be transferred to other language mechanisms (left hemisphere) in order to access the pronunciation or the meaning of written words.

It appears that the right hemisphere is capable of processing certain words, but the results of the studies reviewed here suggest that such processing is easier for words that are short, concrete, imageable, and frequent. Although there is some evidence suggesting that the right hemisphere is capable of attributing a certain degree of meaning to words, at least in terms of an associated image, evidence to the contrary should not be entirely overlooked. Patterson and Besner (1984a) reviewed the studies in which lateralized word presentations were used, in an attempt to uncover any eventual interaction between concreteness (and/or imageability) and visual hemifield, and found that only 6 of 28 experiments reported finding this interaction. In another review of the literature, McMullen and Bryden (1987) noted the presence of this interaction in 6 of 19 experiments involving a reading task and 3 of 7 studies involving a lexical decision task. Finally, Bruyer and Strypstein (1985) found this interaction in 7 of 19 studies where the primary factor was concreteness and in 2 of 7 studies where the degree of imageability seemed to be the primary factor.

As Bradshaw (1980) pointed out, if there is any indication that the right hemisphere has a certain lexical competence, this evidence is often to be found in studies with methodological errors or is based on results that are often difficult if not impossible to replicate. Thus far, no single study has conducted the necessary experiments, carried out the necessary controls, or manipulated all of the pertinent factors to assess, as accurately as possible, the lexical capacities of the normal right hemisphere and/or the nature of the words that it can process.

Finally, the presence of an interaction between the various levels of a given factor (e.g., imageability) and the visual hemifield of presentation is frequently considered important in evaluating the competence of each of the two cerebral hemispheres (see Chapter 2). This interaction suggests that one hemisphere is not solely responsible for word processing. If indeed this were the case, then the differences observed between two levels of a given factor would be the same for both hemispheres. However, Lambert and Beaumont (1981) pointed out that this interaction does not necessarily mean that each hemisphere has its own competence. According to these authors, a loss of information during callosal transfer could amplify any existing hemispheric differences with respect to word processing. However, this controversial hypothesis (e.g., Chiarello, 1989) has yet to be confirmed. A number of other issues also await clarification, thereby reemphasizing the significance of the work that remains to be done.

Semantic Judgments and the Right Hemisphere

The nature of the right hemisphere's lexicosemantic capacities has also been evaluated using semantic judgment tasks. In Day's (1977) experiment subjects were required to decide if a unilaterally presented word (e.g., dog) belonged to the semantic category designated by a superordinate (e.g., animal) which was previously or simultaneously presented in central vision. Reaction times for concrete words (e.g., animals) did not differ as a function of visual-field presentation (e.g., animals). Abstract words (e.g., emotions), however, were recognized much faster when presented in the right visual hemifield. According to Day, the results suggest that the right hemisphere has its own semantic association network and is able to establish semantic relationships, though only when concrete elements are involved. As for abstract words, Day suggests that they may be transferred to the left hemisphere to be processed semantically.

Reaction times obtained by Gross (1972), unlike those of Day (1977), revealed a right visual hemifield advantage for pairs of concrete nouns in a task where the subject had to decide if the two words belonged to the same semantic category. This advantage was independent of the mode of response (i.e., verbal or nonverbal), thereby suggesting that the semantic mechanisms involved were essentially based on the activity of the left hemisphere. To shed some light on the discrepancy between Day's findings and those of Gross, Urcuioli, Klein, and Day (1981) focused their attention on the methodological differences between these two studies. One difference is the mode of response: verbal response and/or manual response involving either proximal muscle control (Gross, 1972) or distal muscle control (Day, 1977). According to Urcuioli et al., the left hemisphere is allowed to control the response in the Gross experiment, in which case the right hemisphere is deprived of any occasion to demonstrate its potential. In Day's experiment, half of the subjects were instructed to respond with a single left-hand finger, a procedure that optimizes the possibilities for the right hemisphere to demonstrate its lexicosemantic competence. A second difference concerns the requirements of the tasks used. In Day's experiment, the task involved a category-inclusion relationship in which the subjects had to decide if a given word belonged to a category of objects represented by a superordinate (hyponymy; category membership task). Subjects in Gross's experiment, however, had to decide if two simultaneously presented words belonged to the same semantic category (cohyponymy; category-matching task).

Urcuioli et al. (1981) repeated the Gross and Day experiments using concrete words and maintaining the mode of response constant: button pushing with a single finger. They found a significant difference between the two visual hemifields, which varied as a function of the tasks and the order in which they were administered. Results showed a right visual hemifield advantage only in the category-matching task, and only when it was presented after the category-membership task. After a certain period of time, it seems that performance on the matching task depends solely on left-hemisphere mechanisms, whereas initially the two hemispheres seem to be equally efficient at performing the category task. Therefore, according to Urcuioli et al., the results of Day (1977) and Gross (1972) are both correct. The apparent contradiction between their findings cannot be attributed to methodological differences. Insisting that there were no significant differences between the visual hemifields for either the membership task or the matching task, when presented first, Urcuioli et al. concluded that both hemispheres were capable of efficiently performing each of the two tasks.

Lambert and Beaumont (1981) used a task in which subjects had to compare pairs of highly imageable words acccording to the size (i.e., imageable attribute) or the pleasantness (i.e., nonimageable attribute) of the objects that were represented by these words. Response was manual, with the hand of response being a manipulated factor. A significant right visual hemifield advantage was found, but there was no interaction between the type of judgment and the visual hemifield of presentation. Lambert and Beaumont concluded that the decision as to which of two imageable or nonimageable words represents the largest or the most pleasant object is made exclusively by the left hemisphere. Indeed, the processes involved in the comparison of word attributes cannot be performed by the right hemisphere. In light of the interaction reported by Day (1979) between imageability and visual hemifield in the context of a lexical decision task, Lambert and Beaumont suggested that the right hemisphere's ability to process imageable words is very limited in right-handed individuals.

In contrast to the findings of Lambert and Beaumont (1983), the results of a study conducted by Rodel, Dudley, and Bourdeau (1983) suggest that the right hemisphere is sensitive to the relationships of synonymy that might exist between imageable words. This was demonstrated using pairs of imageable synonyms (e.g., road–street), nonimageable synonyms (e.g., rage–anger), and semantically unrelated nouns. The first word of each pair was presented in central vision, whereas the second word was presented in lateralized vision. The subjects had to decide if the second word was a synonym of the first. Imageable synonyms were well recognized in either visual hemifield. Results also showed that nonimageable synonyms presented in the left visual hemifield were not recognized as well as imageable and nonimageable synonyms presented in the right visual hemifield. Furthermore, imageable synonyms were recognized better than nonimageable synonyms, but only when they were presented in the left visual hemifield. Thus the right hemisphere seems capable of forming relationships of synonymy between imageable nouns. According to Rodel et al., this capacity whereby two imageable synonyms converge toward the same visual configuration is linked to mental imagery processes for which the right hemisphere is thought to be responsible.

Drews (1984, 1987) also used a semantic judgment task but in a study based on the classic distinction between left- and right-hemisphere functioning, namely, sequential–analytic versus gestalt–holistic. Based on the assumption that these distinctive modes of processing also served as the basis for distinctive semantic organizations, Drews compared intraconceptual relationships with interconceptual relationships (Klix, 1978). Intraconceptual relationships are derived from the analysis of a concept's properties. They lead to a logical classification system and are essentially obtained by way of the analytic processes of the left hemisphere (e.g., bus–train: man-made metal objects, means of transportation). Interconceptual relationships derive from perceived or imagined events of reality. According to Drews (1984, 1987), they depend on the involvement of different concepts in the unity of a scene and can be viewed as the result of the right hemisphere's holistic processes (e.g., coffin–ground: a scene in which a coffin is lowered into the ground). Thus Drews (1984, 1987) compared coordinate relationships (i.e., intraconceptual; e.g., vehicles: bus–train; tools: axe–saw) with locative relationships (i.e., interconceptual; e.g., coffin–earth, shepherd–pasture). Relationships presented in the right visual hemifield were recognized better. Drews (1987) attributed this finding to the left hemisphere's superiority for the processing of linguistic material. However, intraconceptual relationships were understood better than interconceptual relationships when they were presented in the right visual hemifield, whereas the opposite was true for stimuli presented in the left visual hemifield. Moreover, Drews (1987) reported that in comparison to pairs of semantically unrelated words, locative relationships were significantly easier to process when they were presented in the left visual hemifield, whereas coordinate relationships were significantly easier to process when they were presented in the right visual hemifield. The processing of intraconceptual relationships therefore appears to be more efficient in the left hemisphere, whereas the processing of interconceptual relationships is seemingly more efficient in the right hemisphere (Drews, 1984, 1987). According to Drews (1987), the semantic organization sustained by the right hemisphere is qualitatively different from that sustained by the left hemisphere: each semantic structure (i.e., intraconceptual versus interconceptual) is exclusive to one hemisphere or the other.

Recently, Rodel, Landis, and Regard (1989) reported the results of a study also suggesting that each hemisphere sustains qualitatively different lexicosemantic organization, or that each hemisphere processes word meaning differently. This study was based on a preliminary analysis of the semantic paralexias produced by normal subjects when words were projected very briefly in one or the other hemifield (Regard & Landis, 1984; also see Chapter 5). According to this analysis, the right hemisphere seems to more frequently produce paralexias that are semantically distant from the target word, whereas the left hemisphere produces what the authors termed "classical semantic paralexias" (i.e., semantically close to the target words).

To further investigate this difference, Rodel et al. (1989) undertook two experiments in which pairs of words were projected tachistoscopically. In each experiment, and for each pair of stimuli, one word was projected in the center of the visual field while the other word was projected in the left or right visual field. Subjects were asked to indicate, for each pair, whether they felt that the two words were related in meaning. In the first experiment pairs of semantically related words were selected on the basis of their predetermined semantic distance: close[4] vs. distant (e.g., autumn–age). In the second experiment 28 word pairs were created from 8 nouns randomly chosen. These pairs were projected in lateralized vision in the context of a semantic judgment task and then rated by each subject in free vision according to their degree of semantic relationship (close, medium, distant, and no semantic relation). The same pattern of results was found in the two experiments whether the nature of the semantic relation had been predetermined or rated a posteriori: pairs were considered to be related more frequently in the right hemisphere than in the left hemisphere when the stimuli involved remote or no semantic relations, whereas the converse was true for the word pairs involving close or medium semantic relation. Rodel et al. concluded that a functional dissociation exists between the two hemispheres with respect to their estimation of semantic relationships. According to these authors, the two hemispheres appreciate word meaning differently and probably sustain different reading strategies.

Despite the interesting nature of the finding of Rodel et al., the results of this study need some clarification. For instance, the paradoxical sensitivity of the right hemisphere for the presence of a semantic relation when no relation exists (experiment 2) remains to be explained. Indeed, it seems amazing that unrelated words, at least as rated in free vision, would be considered as semantically related in a semantic field sustained by the right hemisphere. The nature of the word pairs involving a distant semantic relation also needs elucidation. Nevertheless, this study suggests a tendency that should be further investigated.

In conclusion, the results from semantic judgment tasks show that the normal right hemisphere can sometimes demonstrate a certain capacity to process words

[4]Derived from a preliminary experiment and consistent with lists of association frequency. Rodel et al. (1989) provided no example of "closely" related semantic pairs.

semantically. The normal right hemisphere appears to be sensitive to different semantic relationships (e.g., subordinate relationships, category inclusion, and synonymy), at least when concrete and/or imageable elements are processed. Moreover, certain semantic relationships seem to be given preferential treatment by the right hemisphere (Drews, 1984, 1987; Rodel et al., 1989). For instance, the relationships between objects belonging to a perceived event or including relationships based on contiguity seem to be more readily accessible to the right hemisphere than are cohyponymic relations. Furthermore, the right hemisphere would be more sensitive to remote semantic relations than to semantic relationships that are close, and more sensitive than the left hemisphere to relationships that are rated as distant. However, the exact role of the contiguity factor and of the semantic distance should be investigated in the context of an orthogonal analysis of these two factors. At the time being, it is interesting to note that the contiguity effect and the semantic distance effect are compatible with the preponderance of syntagmatic errors made by the disconnected right hemisphere of split-brain patients and with the semantic paralexias produced by the right hemisphere in normal subjects (cf. Regard & Landis, 1984).

Semantic Interference

The literature concerning the right hemisphere's contribution to lexical semantics includes a large number of studies based on the general principle of semantic interference. Three types of semantic interference are examined here: Stroop tasks, modified Stroop tasks, and priming tasks.

Stroop Studies

The Stroop effect (Stroop, 1935) concerns the interference that occurs when a subject is required to name the color of the ink used to print a color word when the word differs from the color of the ink. This type of interference is at work, for example, when a subject says that the word "red" is printed in red when it is printed in green. In a task such as this, conflicting information results in semantic interference during stimulus processing. The amount of interference is evaluated by the number of errors produced and/or by an increase in the amount of time needed to name the color of the ink.

Dyer and Harker were the first to use the Stroop test in visual laterality studies (Tsao, Feuster, & Soseos, 1979). Reaction time data from the two experiments conducted by these investigators showed no significant difference between the left and right visual hemifields (Dyer, 1973). Tsao et al. reported similar results, even after having increased the visual angle between the stimulus and the point of fixation. However, an analysis of error frequency revealed fewer errors for stimuli presented in the left visual hemifield, thereby suggesting that the Stroop effect had a greater influence on left-hemisphere processes (Tsao et al. 1979).

In another experiment, the color words were presented horizontally (Hugdahl & Franzon, 1985) rather than vertically, as in the Tsao et al. (1979) study. The results (i.e., reaction time and error frequency) revealed a greater Stroop effect for color-word stimuli presented in the right visual hemifield of right-handed male subjects. According to Hugdahl and Franzon, this left hemisphere dominance for the semantic processing of incongruent color-word stimuli is all the more clear if one considers that in the absence of interference, there is no significant difference between the visual hemifields for color naming. These investigators concluded that this dominance reflected the actual functional asymmetry of the brain for language since they also observed that the right hemisphere of left-handed males was more sensitive than the left to the Stroop effect.

To extend their observations concerning the lateralization of Stroop effects in the intact brain, Franzon and Hugdahl (1986) used a vertical display of incongruent color-word stimuli. Their results showed a significantly greater interference in the right visual hemifield of right-handed subjects when reaction times were considered, and particularly in right-handed males when error frequency was considered. A similar trend was observed in left-handed males, but no difference was found in right- or left-handed females. According to Franzon and Hugdahl, these results confirm the left hemisphere's dominance for language processing in the right-handed male and support the notion of greater bilateral functioning in females.

A rather general statement can be made here concerning the aforementioned studies. Indeed, by requiring subjects to orally name the color of the ink in which a color word is printed, these studies create a bias that is unfavorable to the manifestation of the lexicosemantic potential of the right hemisphere. However, Schmit and Davis (1974) designed an experiment that overcame this limitation by using a manual response, whereby subjects had to press one of three buttons (red, green, or blue). The subjects, all right-handed males, had to identify either the color of the ink or the name of the color word. Half the subjects responded with their right hand; the other half responded with their left hand. There were two types of stimulus presentations: color words were printed either with incongruent colors (Stroop stimuli) or with congruent colors (non–Stroop stimuli). The color-word stimuli were presented in either the left or right visual hemifield. The analysis of reaction times revealed an interaction between task (ink-color identification, word-color identification) and the hemisphere to which the stimuli were presented. This interaction suggests that the left hemisphere was slower than the right in selecting the color of the ink but was relatively faster in identifying the color name presented. Furthermore, when selecting the color of the ink, the two hemispheres were equally efficient when processing non–Stroop stimuli, but the left hemisphere was slower than the right when processing Stroop stimuli. The Stroop effect is therefore more important in the left hemisphere, though it is also observed in the right hemisphere.

Schmit and Davis (1974) also reported a significant interaction between stimulus type (Stroop, non-Stroop), hemisphere of stimulus presentation, and response hand. Overall, the left hand responded faster than the right. In the case of the right hemisphere, however, the difference between Stroop and non–Stroop stimuli was greater when the right hand was used (66 msec) but negligible when the left hand was used (13 msec). As for the stimuli presented to the left hemisphere, this difference remained constant regardless of which hand was used to respond. According to Schmit and Davis, the Stroop effect does not occur unless the left hemisphere is involved in stimulus reception or analysis or in the organization of the response. Only the left hemisphere seems capable of sustaining lexicosemantic processes.

The results of color-naming tasks support this conclusion. Indeed, they show (1) that the left hemisphere is generally faster than the right, (2) that the right hand responds faster than the left in all conditions, and (3) that there is no interaction between stimulus type (i.e., Stroop, non-Stroop) and the hemisphere to which the stimuli are presented, although individually each of these factors plays a significant role. In summary, the findings of Schmit and Davis (1974), as well as those of others who have studied the Stroop effect, confirm the classic superiority of the left hemisphere for the semantic processing of words in the right-handed individual. They provide few if any arguments in favor of the right hemisphere's contribution to lexical semantics.

Modified Stroop Studies

Several authors have modified the incongruent color-words paradigm in order to create a different type of semantic interference. These modified versions of the Stroop test are also based on a semantic incongruence, but they involve pictures of objects rather than color words. For example, Wuillemin, Krane, and Richardson (1982) devised a task in which the Stroop effect was dependent on the interference created by the presentation of a word (e.g., dog) during picture naming (e.g., lion) when a semantic relationship existed between the word and the picture. The pictures were presented either individually or simultaneously along with a name designating an object belonging to the same semantic category as the object in the picture (e.g., animal or food). When the word and the picture were simultaneously presented in the same visual hemifield, the word was superimposed over the picture. The subjects were instructed to ignore these words. Few errors (2.4%) were made on this task and there were no significant differences (Wuillemin et al., 1982). In contrast, reaction times revealed a Stroop effect in both visual hemifields when the word and the picture were presented ipsilaterally. Despite absence of any visual field differences, Wuillemin et al. concluded that there was no semantic interference in the right hemisphere, a conclusion that is more the reflection of the authors' belief in the exclusiveness of the left hemisphere to word processing than the reflection of their observation.

In another study, Lupker and Sanders (1982) instructed their subjects to name laterally presented pictures. In one experiment, there were three different types of presentation: (1) the picture was presented alone; (2) a semantically related word was superimposed on the picture (e.g., apple and banana); and (3) an unrelated word was superimposed on the picture (e.g., piano and banana). Results showed that subjects made few errors and that there was no apparent trend. With respect to reaction times, the amount of interference introduced by the presence of the word was calculated by subtracting the amount of time required to name pictures only from the amount of time required to name pictures with superimposed words. The interference effect was illustrated by an increase in reaction time when a word was superimposed on a picture. The analysis of this interference reveals two important points. On the one hand, when there is a semantic relationship between the picture and the distracting word, the resulting interference is statistically more important in the right than in the left visual hemifield. This led Lupker and Sanders to hypothesize that words presented to the right hemisphere are not as fully processed as those presented to the left hemisphere and are therefore less likely to interfere with the processing of the picture. On the other hand, when an unrelated word was presented with the picture, the resulting interference was the same in both visual hemifields. Thus it is only in the presence of a semantic relationship that the interference is more important in the right than in the left visual hemifield. According to Lupker and Sanders, one is therefore tempted to conclude that words presented to either of the cerebral hemispheres receive similar phonetic processing, as suggested by the interference produced in the absence of a semantic relationship between word and picture, whereas only words presented to the left hemisphere receive adequate semantic processing.

In a second experiment, Lupkers and Sanders (1982) sought to verify their hypothesis. In addition to the three types of stimulus presentation described above, two other types were also used: a picture with a pronounceable nonword (e.g., a picture of a lion with the nonword "dera") and a picture with a series of consonants (e.g., a picture of a dog and the series "scnrv"). Furthermore, at the end of the task, the subjects were asked to recall all of the words presented (i.e., memory task). Lupker and Sanders observed that with respect to stimuli presented in the right visual hemifield, the interference effect had a phonetic as well as a semantic component. Conversely, the interference effect associated with stimuli presented in the left visual hemifield was independent of the linguistic nature of the letters superimposed on the picture. Furthermore, during free recall, fewer words were reported from the left visual hemifield than from the right visual hemifield, whereas the number of words recalled was greater for words that were semantically related to the picture than for words that were not, whatever the hemifield. In light of these results, Lupker and Sanders concluded that words presented in the left visual hemifield were processed semantically but not phonetically. Thus the apparent absence of semantic interference in the picture-naming task, when stimuli are presented to the right hemisphere, is due in reality to the absence of the necessary phonetic processing; both types of processing are essential for semantic interference to occur. According to Lupker and Sanders, the right hemisphere is therefore capable of overseeing the semantic processing of words.

The potential interference effect resulting from the simultaneous presentation of a picture and its corresponding name has also been investigated by Underwood (1976, experiment 2). However, the results reported by this author show that the amount of time required to name the picture of an object is not influenced in any specific way by the simultaneous presentation of the object's name, regardless of the visual hemifield involved, when the attention of the subjects is not spatially directed. In addition to the object's name, Underwood used other words that were semantically related or not to the pictures. He observed that for the left hemisphere, the inhibition induced by words that were unrelated to the target was significantly greater than when words were semantically related to the target, whereas for the right hemisphere this comparison was not significant. Furthermore, the presence of words semantically related or not to the target resulted in a slowing of picture naming that was more important for stimuli presented to the left hemisphere. Underwood interpreted this inhibition in terms of an easier access to the lexicon and/or the language production system and situated this access in the left hemisphere for the right-handed individual. Thus little place is given to the right hemisphere's capacity for lexicosemantic processing.

In another experiment conducted by Underwood (1977), subjects' attention was spatially directed in the context of a Stroop-like picture-naming task. The subjects were required to name the picture but also to report the name of the word accompanying each picture. The purpose of this last requirement was to test for awareness of the irrelevant words. Data analysis revealed a three-way interaction between the following variables: the visual field in which the interfering word was presented, the presence of a semantic relationship between the word and the picture, and word report (i.e., the subjects' awareness of the distracting stimuli). Part of this interaction was related to the fact that the interference caused by words semantically related to the pictures was restricted to right visual hemifield presentations when the subjects were unaware of the distracting words (i.e., unreported instances). Underwood concluded that the lexicon was specific to the left hemisphere in right-handers.

Since picture naming creates a bias favoring the left hemisphere, Underwood and Whitfield (1985, experiment 2) attempted to avoid this bias by asking their subjects to simply state whether the picture represented an animal. The interference anticipated in this semantic judgment task therefore occurs at the level of picture categorization. The distracting elements were semantically related or unrelated words and pronounceable nonwords. When a semantic relationship did exist, it was one of two types depending on whether the word was an animal name related to the picture (e.g., tiger for the picture of a swan) or a word related to a more general meaning of the picture (e.g., hunt for the picture of a fox). The distracting element and the picture were simultaneously presented in either of the two visual hemifields, with the distracting element positioned either above or below the picture. The subjects were instructed to disregard the distracting element. Finally, a visual mask consisting of letter fragments immediately preceded and followed each stimulus presentation.

Underwood and Whitfield (1985, experiment 2) reported an interference effect (i.e., a facilitating effect) limited to trials in which the distracting element was related to the general meaning of the picture and when both elements were presented in the left visual hemifield. According to Underwood and Whitfield, this suggests that the right hemisphere does have a certain potential for semantic processing. However, these investigators were aware of the fact that their findings contradicted those of Lupker and Sanders (1982). One possible explanation for this difference concerns the tasks used in these two studies: judging if a word belongs to the animal world versus picture naming. A second possible explanation concerns the use of a visual mask in the Underwood and Whitfield experiment but not in the Lupker and Sanders experiment. Not only does the visual mask make stimulus recognition more difficult, but it can also reverse the visual hemifield advantage.

Underwood and Whitfield (1985, experiment 3) sought to verify this latter hypothesis. Indeed, by manipulating the presence or absence of a visual mask, Underwood and Whitfield were able to confirm their earlier results. However, in the case of right-hemisphere presentations, the facilitating effect produced by the presence of a word related to the meaning of the picture when visual masking was present became an inhibiting effect when there was no such masking. Furthermore, in the absence of visual masking, the presence of a semantic relationship does not regulate the interference induced by the presence of a word when the word and the picture are presented to the left hemisphere, a result that is in sharp contrast with the findings of Lupker and Sanders (1982). Finally, the right hemisphere advantage for picture categorization was not observed in the presence of visual masking. According to Underwood and Whitfield, it is quite likely that picture categorization processes are sustained by the right hemisphere and that this hemisphere is also capable of processing words semantically. As a result, it is normal that a word–picture interaction be observed only for the right hemisphere. Indeed, there is a facilitating effect when the processing of a word helps the identification or the categorization of a picture by providing certain clues about the nature of the object to be processed. Conversely, inhibition occurs when the word is processed to the extent that it interferes with and blocks picture processing. This is what happens in the situation where there is no masking, since the word provides an immediate response that can interfere with the processing of the picture.

In summary, of the five studies reviewed here, only two suggest some degree of a right-hemisphere–based lexicosemantic potential, and one of those reported contradictory findings (i.e., Lupker & Sanders, 1982). The difficulty in reaching any unanimous conclusion concerning the true lexicosemantic potential of the right hemisphere can be attributed, at least in part, to methodological differences and limitations. It is possible, for example, that the nature of the tasks is favorable (e.g., tasks using pictures) or not (e.g., naming tasks) to the manifestation of lexicosemantic processes controlled by the right hemisphere.

Semantic Priming

Semantic priming corresponds to a condition in which recognition of a target is facilitated by way of the previous or simultaneous presentation of a prime referring to one or more indications concerning the identity of the target. Therefore, unlike the Stroop paradigm, semantic priming is based on a mutual facilitating interference between two items, rather than on the presence of a semantic incongruity that has to be resolved. An example of semantic priming in a lexical decision task would be to present the prime word "nurse" followed by the target word "doctor." The activation of the semantic field related to the word "nurse" should facilitate the recognition of the word "doctor" as being a part of the English language.

To evaluate the relationship between word imageability and the functional lateralization of the brain for language, Deloche Seron, Scius, and Segui (1987) controlled the orthographic features of their word stimuli. Thus some of the target words were homographs[5] having two different meanings, one of a high-imagery value and the other of a low-imagery value. Each homograph was associated with two primes that were semantically related either to the high-imagery meaning or to the low-imagery meaning. Nonhomograph words, similar to the homographs in terms of their frequency, were also used as targets. Half of these targets were associated with a prime of high-imagery value; the other half were associated with a low-imagery prime. The primes were presented in central vision, followed by the presentation of a target either in the left or in the right visual hemifield. A lexical decision task involving a bimanual response was used. The total number of errors was very small (20 of 4026 responses). The presence of a semantic relationship between nonhomograph words produced shorter decision times when the primes were presented in the right visual hemifield or when they were of high-imagery value. However, there was no significant interaction between these two factors. When the targets were homographs, the presence of a semantic relationship produced an interaction suggesting that targets presented in the right visual hemifield were processed faster. However, this finding reached statistical significance only when the primes were associated with the low-imagery meanings of the homographs. The nature of the prime had no significant influence on the processing of homographs in the right visual hemifield. As far as left visual hemifield presentations were concerned, homographs preceded by a prime associated with their high-imagery meaning were processed more rapidly than those preceded by a prime associated with their low-imagery meaning. On the basis of their findings, Deloche et al. concluded that the right hemisphere can control the activation of highly imageable meanings only when processing ambiguous words such as homographs. Since the subjects in this study were all female, it remains to be seen whether this conclusion can be generalized to male subjects.

In another study of semantic priming innvolving a lateralized lexical decision task, Zaidel (1983a, 1986b) used concrete and frequent imageable words. In this experiment, the prime and the target were presented either ipsilaterally or contralaterally. The subjects responded by pushing a key with their right hand whenever the target was a word of the English language. They were instructed to ignore the prime and to focus their attention on the fixation point. No statistical

[5]The term "homograph" refers to orthographically similar words that have different meanings.

tests were reported, but the difference in reaction times between the trials with a semantically related prime and those without primes (facilitative effect), on the one hand, and between trials with a semantically unrelated prime and those without primes (inhibitory effect), on the other hand, indicate that the right hemisphere is sensitive to semantic priming. Zaidel (1983a, 1986b) adds that the primes produce no inhibition and little facilitation with respect to left-hemisphere presentations, whereas they result in stronger inhibition and greater facilitation with respect to right-hemisphere presentations. The automatic activation of lexicosemantic organization therefore appears to be greater in the right hemisphere than in the left (Zaidel, 1983a, 1986b). According to Zaidel (1986b), this observation is in agreement with the existence of independent lexicosemantic networks that are differently organized within each hemisphere.

Although Zaidel (1986b) considered his priming condition to be of the automatic type, the methodology he used does not guarantee the automatic nature of semantic priming (Chiarello, 1986), nor does a similar guarantee exist in the study conducted by Deloche et al. (1987). Among other things, the duration of the primes (i.e., Zaidel, 100 msec; Deloche et al., 500 msec), the proportion of stimulus pairs containing a semantic relationship (i.e., Zaidel, 50% of the trials in which the target was a word; Deloche et al., 60%), as well as the prime– target interval (i.e., Zaidel, 500 msec; Deloche et al., 1 sec) may have favored a controlled use of semantic relationships. Indeed, under such priming conditions it is possible for the subjects to elaborate certain strategies and anticipate the nature of the targets in order to improve their performance (see below). In fact, the presence of inhibition, as in the Zaidel study (1986b), usually attests to the conscious use of primes, which results in incorrect expectations (Neely, 1977; Posner & Snyder, 1975).[6] Therefore, the automatic nature of the priming effect observed in these studies remains questionable.

Marcel and Patterson (1978) examined automatic semantic priming using a task in which the presentation of each prime was followed by visual masking. This procedure enhances the automatic nature of the semantic priming. Indeed, the mask makes it more difficult to recognize the primes and, as a result, conscious use of the prime becomes more difficult. In this study, the primes were either of high-imagery value or low-imagery value; all of the targets were imageable. Results showed that the degree of imageability had no influence on the priming effect. Marcel and Patterson also noted the absence of any significant difference between the two visual hemifields, regardless of whether the primes were presented ipsilaterally or contralaterally to the targets. However, the amount of priming was significantly greater for ipsilateral prime–target presentations. Marcel and Patterson therefore concluded that semantic processing of isolated words can be accomplished by either hemisphere.

[6]One might note, however, that this point is still debatable (see McLeod & Walley, 1988).

Underwood, Rusted, and Thwaites (1983) introduced the use of homophone[7] primes in examining the effects of semantic priming in a lexical decision task. The prime could be either a word that was semantically related to the target (e.g., sweet–bitter) or a word that was orthographically different from but pronounced like a word that was semantically related to the target (e.g., suite–bitter). Performance for these two types of prime–target pairs was compared to the performance for other pairs in which the homophone prime was not semantically related to the target (e.g., hair–tower). By using homophones it is possible to determine whether phonological processing precedes the semantic processing of the primes. For example, if the word "suite" is phonologically processed, after undergoing a grapheme–phoneme conversion it could be interpreted as "sweet" and therefore prime to the word "bitter."

In addition to the prime–target pairs just described, Underwood et al. also used two other types of stimuli: nonhomophonic primes that were either semantically related (e.g., plate–dish) or nonrelated (e.g., smile–thing) to the targets. The targets were presented in central vision while the primes were simultaneously presented in either visual hemifield. The automatic nature of this priming task was ensured by the brief presentation of each prime–target pair (i.e., 50 msec) as well as by the presence of a visual mask immediately preceding and following each presentation. Moreover, the subjects were told that only the centrally presented word was relevant. The laterally presented word was characterized as a distractor that was to be ignored. The subjects had to indicate if the target was a word or a nonword by pushing one of two keys. Half of the subjects responded with their right hand while the other half used their left hand. Reaction times and accuracy scores revealed that performance was influenced by the nature of the semantic relationship between the primes and the targets. On the one hand, the lexical decision response was more difficult when the nonhomophone primes were semantically related to the targets as opposed to the situations in which the primes were semantically unrelated to the targets. On the other hand, the lexical decision response was more difficult when the homophone primes were directly or indirectly semantically related to the targets as opposed to the situations in which the homophone primes were semantically unrelated to the targets. There was no significant difference between the two hemispheres.

The priming effect reported by Underwood et al. (1983) appears to stem from an inhibitory effect on target processing, a result that is not usually expected, or desirable, when one is supposedly engaged in automatic priming (Neely, 1977; Posner & Snyder, 1975; but see McLeod & Walley, 1988). Despite this, Underwood et al. concluded that automatic, unconscious semantic processing of isolated words is possible in either hemisphere. Moreover, according to these investigators, the priming effect obtained with homophones suggests that grapheme–phoneme conversion is not exclusive to the left hemisphere.

---

[7]The term "homophone" here refers to phonologically similar but orthographically dissimilar words that have different meanings.

Walker and Ceci (1985) used a rather different lexical decision task with semantic priming. In this task, the primes and targets were simultaneously presented (100 msec) in the right, left, or central visual field. The subjects were instructed to push either a response key labeled "yes" if both items of the prime–target pair were actual words or a response key labeled "no" if one of the items was not an actual word. The subjects responded with the right hand on half of the trials and with the left hand on the other half. A priming effect was expected to be found for syntactically related word pairs (e.g., dog–bark) and for categorically related pairs (e.g., rain–snow). Other stimuli included unrelated word pairs (e.g., dog–rain) and word–nonword pairs (e.g., ant–trem). The analysis of response times revealed no significant difference between the right and left visual hemifields in terms of priming effect. It seems therefore that the semantic component of words is automatically processed regardless of the visual hemifield involved. The results also showed that the proportion of correct responses for syntactically related pairs was superior to the performance for unrelated pairs in both visual hemifields. A similar comparison involving categorically related pairs revealed a difference only for stimuli presented in the left visual hemifield. This finding led Walker and Ceci to speculate that sensitivity to priming varies from one visual hemifield to the other, but only for certain types of primes (i.e., those with categorical relationships). However, Walker and Ceci were skeptical of the automatic semantic processing involved since their task required that subjects pay attention to both items of each pair. Further, they were unsure whether semantic priming effects were restricted to situations in which conscious semantic judgments were not required (e.g., lexical decision tasks) or if such effects could also be observed when the dependent variable was a qualitative index of the subject's semantic judgment.

To answer these questions, Walker and Ceci (1985) conducted a second experiment in which the unconscious nature of semantic activation was achieved by making it impossible for the subjects to identify the primes. Furthermore, because the subjects were required to interpret homographs, the cognitive requirements were somewhat more important than those of a standard lexical decision task. In this experiment, the priming effect was dependent on the influence that a word could have on the processing of a target which had two different meanings. The homographs were presented in central vision and were flanked by two words, one presented in the left visual hemifield and the other in the right visual hemifield. Each homograph (e.g., ball) had a dominant meaning (e.g., a round sporting object) and a nondominant meaning (e.g., a formal dance). On some of the trials, the pair of flank words were unrelated to the homograph. On other trails, one of the two flank words was related to the nondominant meaning. Immediately following the stimulus presentation (3 ms), a visual mask was presented in order to prevent the identification of the words flanking the homograph. The subjects had to respond by giving a word association or a definition of the homograph. Walter and Ceci found that primes semantically related to the nondominant meaning of the homographs exerted a similar influence in both hemifields. It is obvious, according to these authors, that the semantic attributes of words can be automatically (unconsciously) taken into consideration when stimuli are presented in either visual hemifield.

According to Chiarello (1983a, 1983b, 1985), the degree of consciousness or control that is needed to process semantic relationships modulates the functional lateralization of the brain. Recalling the findings of Milberg and Blumstein (1981), Chiarello (1983a, 1985) postulated that the preservation of automatic semantic processes in aphasia was due to the intact right hemisphere. However, the conscious use of semantic knowledge, which was shown to be disturbed by left-hemisphere damage, was specifically dependent on the left hemisphere. Chiarello (1983a) therefore hypothesized that automatic semantic processing is performed just as efficiently when the stimuli are presented in either visual hemifield, whereas conscious or active semantic processing is performed more efficiently when the stimuli are presented in the right visual hemifield. In an attempt to verify this hypothesis, Chiarello (1983a, 1983b, 1985) conducted a series of lexical decision experiments involving both automatic and conscious or controlled priming.

Chiarello (1983a, 1985) used the terms automatic priming and controlled priming in their most widely accepted meaning. Thus automatic priming is believed to be based on the existence of semantic networks in memory. The priming facilitation is thought to result from node-to-node semantic activation (cf. "spreading-activation theory"; Collings & Loftus, 1975). Controlled priming, however, requires more than just the existence of semantic networks. It results from the active, conscious, and voluntary use of the prime in order to identify which components of the semantic network are the most relevant in forming some sort of expectation with respect to the target.

Controlled priming is associated with benefits (e.g., faster reaction times in lexical decision) given (1) the existence of a semantic relationship between the prime and the target, and (2) a conscious expectation as to the semantic nature of the target. However, controlled priming also results in costs (e.g., slower reaction times in lexical decision) in those instances in which the target is unrelated to the prime, given an erroneous conscious expectation as to the semantic nature of the prime. Automatic priming, on the other hand, results in benefits only when the target and the prime are semantically related and in no costs when the target and the prime are unrelated. Indeed, in this priming situation, the prime does not lead to the formulation of any expections.

Chiarello (1983a, 1985) used concrete, familiar, and imageable nouns to examine automatic and controlled priming in a lexical decision task. The proportion of prime–target stimulus pairs containing a semantic relationship (e.g., dog–cat) and the instructions given to the subjects were varied in the two conditions. Thus in the controlled priming condition, semantically related pairs represented 75% of the pairs in which the target was a word. The subjects were instructed to pay attention to the meaning of the prime because it would facilitate the recognition of the upcoming target. Conversely, in the automatic priming condition, only 25% of the word pairs contained a semantic relationship. The subjects were told that the prime served only as a warning signal for the target stimulus. The purpose of this procedure was to prevent the subjects from using the prime as a semantic predictor of upcoming target words. In both conditions, the prime (100 msec) and the target (300 msec) were presented successively (interstimulus interval: 500 msec) and ipsilaterally. The subjects had to respond by pushing a key with the right index finger (word) or middle finger (nonword).

Chiarello (1983a, 1985) observed that decision making was both faster and more accurate for stimuli presented in the right visual hemifield. This finding is a reflection of the traditional dominance of the left hemisphere for language. As for the influence of priming effects, which correspond to the combined effects of costs and benefits, it was found that active or controlled semantic priming was more important in the right visual hemifield, whereas automatic semantic priming was more important in the left visual hemifield. These results led Chiarello (1983a, 1985) to conclude that automatic semantic activation is not a specialization of the left hemisphere, which explains why it is not disturbed in the aphasic patient. The left hemisphere oversees only conscious semantic activity, under the voluntary control of the subject.

Interestingly, the automatic-controlled dissociation reported for the semantic processing of concrete words has not been consistently observed by Chiarello. In a study on the effects of aging, for example, Chiarello, Church, and Hoyer (1985) reported that the controlled priming effect was more important in the right visual hemifield in women only. No interhemifield difference was reported with respect to automatic priming or controlled priming in men or with respect to automatic priming in women. In a more recent study, Chiarello and colleagues (Chiarello, 1986; Chiarello, Senehi, & Nuding, 1987) used abstract (e.g., hunger) and concrete primes (e.g., bread) paired with concrete targets (e.g., food). The primes were presented in central vision, whereas the targets were presented in either the left or the right visual hemifield. The results showed that the strength of the priming effect was significantly different in the two visual hemifields but only when priming was controlled and when the primes were abstract words. In addition, the priming effect induced by the concrete primes was stronger than the effect resulting from abstract primes, but only when the stimuli were presented in the left visual hemifield. Finally, in the controlled priming condition, the priming effect observed for stimuli presented in the right visual hemifield resulted from benefits and significant costs, whereas the priming effect for stimuli presented in the left visual hemifield resulted only from benefits (no costs). According to Chiarello et al. (1987; Chiarello, 1986), these findings point to the existence of interhemispheric differences related to the postlexical processing of abstract words, that is, once the lexicon has been accessed and the meaning of the prime has been integrated. Abstract or nonimageable words presented to the right hemisphere appear to be the object of limited postlexical processing, the right hemisphere being less capable than the left for this type of processing. However, the right hemisphere appears to be just as capable as the left of automatically accessing the lexical representations of concrete and abstract words and of controlling the postlexical processing of concrete words. Thus it would be false to assume that the right hemisphere has access only to a lexicosemantic network that is restricted to concrete and highly imageable words (Chiarello et al., 1987).

According to Chiarello (1986, 1988a), the automatic activation of semantic networks is possible in each cerebral hemisphere, but only the left hemisphere is capable of actively using semantic knowledge to facilitate the processing of certain items and inhibit the processing of others. Whereas the left hemisphere is capable of focusing on a single element of a semantic network, the right hemisphere maintains several active elements of this network. The inhibitory effect that is specific to the left hemisphere could also explain why the Stroop effect is more important for stimuli presented to this hemisphere. In support of this hypothesis, Chiarello (1988a) emphasizes the fact that the controlled priming effect for stimuli presented to the right hemisphere is associated only with benefits (no costs), whereas the same priming effect for stimuli presented to the left hemisphere is associated with both benefits and costs.

The results of a lexical priming experiment conducted by Burgess and Simpson (1988) support Chiarello's (1986, 1988a) hypothesis. In this experiment, the primes were ambiguous words (homographs) presented in the central visual field while the targets were semantically related words presented in the left or right visual hemifield. The duration of the prime–target interval was varied [SOA (Stimulus Onset Asynchnony): 35 msec vs. 750 msec] as was the degree of association between the primes and the semantically related targets. Burgess and Simpson observed that when the targets were projected to the left visual hemifield and at the shorter SOA, the ambiguous words primed only the dominant meaning. At the longer SOA, however, both meanings were activated. In contrast, when the targets were presented in the right visual hemifield and at the shorter SOA, the subordinate meaning as well as the dominant meaning was activated. At the longer SOA, a facilitation effect was observed for the dominant meaning while the subordinate meaning was inhibited (related targets produced slower responses than unrelated targets). Burgess and Simpson concluded that this inhibition effect supported Chiarello's (1985) finding that controlled semantic processing is specific to the left hemisphere, whereas automatic processing occurs in both hemispheres. Under normal conditions, it seems that the left hemisphere is more selective, focusing on one meaning only and inhibiting all others, while the right hemisphere retains all possible meanings of a word for future use in the task or by the left hemisphere (Chiarello, 1986, 1988a; Burgess & Simpson, 1988).

The final point to be examined here concerns the right hemisphere's superiority for automatic priming. This superiority was reported by Chiarello (1983a, 1985) and recently confirmed by Michimata (1987, cited by Chiarello, 1988b). In an attempt to compare the effects of automatic and controlled priming, Michimata varied the prime–target onset interval. At the shorter interval (200 msec) Michimata observed that priming effects were more important for targets presented in the left visual hemifield, whereas at the longer interval (1500 msec) the priming effects were more important for right visual hemifield presentations. In her explanation of the right hemisphere's superiority for automatic semantic priming,

Chiarello (1983a, 1985, 1988) refers initially to the nature of the lexicon to which this hemisphere has access. Postulating that the amount of semantic activation is inversely proportional to the size of the semantic network, Chiarello (1983a) suggested that the right hemisphere's lexicon is a subset that is either independent of or shared with the left hemisphere's lexicon. The words included in this subset are concrete and imageable, and they have a high frequency of use.

As Chiarello (1988b) recently noted, this superiority of the right hemisphere has not always been reported and there are some studies in which the effects of automatic priming have been found to be similar in both visual hemifields (e.g., Burgess & Simpson, 1988; Chiarello et al., 1987; Walker & Ceci, 1985). To explain these contradictory findings, Chiarello (1988b) singled out two methodological details in the studies that have reported greater automatic priming effects in the left visual field. These are the central versus lateral presentation of the primes and the nature of the semantically related prime–target pairs. Indeed, according to Chiarello (1988b), the words forming these pairs are semantically similar (e.g., king–duke), though they are not necessarily associated. Studies that have failed to find the right-hemisphere superiority used pairs of words that were associated though not semantically similar (e.g., police–jail). In a recent automatic priming experiment, Chiarello compared the location of the primes (central versus lateral) and the nature of the prime–target semantic relationship (similar but not associated: e.g., deer–pony; similar and associated: e.g., boot–shoe; associated but not similar semantically: e.g., bee–honey). She found that when the primes were presented in central vision, the priming effect was similar in both hemispheres regardless of the nature of the prime–target relationship. When the primes were laterally presented, a priming effect was observed for "similar but not associated" pairs presented in the left visual hemifield. No hemifield differences in the size of the priming effect were observed for lateral presentations of "similar and associated" pairs. No priming was obtained with "associated but not similar" pairs when primes were lateralized. According to Chiarello, these results suggest that automatic semantic activation (spreading activation) is similarly induced in both hemispheres when highly related words (similar and associated) are involved. They also suggest that the right hemisphere plays a particular role in the processing of semantically similar concepts or of word-pairs with a low degree of semantic overlap (Chiarello, 1988b).

In summary, a number of lateralized semantic priming studies have shown that the right hemisphere is capable of sustaining automatic semantic or lexicosemantic activity. The right hemisphere's semantic activity can be triggered by concrete and/or imageable words and probably by abstract and/or nonimageable words. Moreover, this activity appears to be of greater importance for certain types of semantic relationships, namely, distant ones. As for the right hemisphere's capacity for active, conscious, voluntary, or controlled processing of concrete words, the results from priming experiments suggest that this capacity is still questionable. However, this capacity appears to be extremely reduced if not totally absent with regard to abstract words and their semantic relationships.

Conclusion

The results of divided visual field experiments with normal right-handed subjects indicate that the existence of a lexicosemantic potential sustained by the right hemisphere is still a controversial issue with respect to the sensitivity of the right hemisphere to the concrete and/or imageable nature of words. In spite of this controversy, studies involving semantic judgments suggest that the right hemisphere is sensitive to different types of semantic relationships. Studies using the classic version of the Stroop test have shown a greater sensitivity of the left hemisphere to the semantic component of the lexicon, whereas modified versions of the Stroop paradigm are more favorable to the right hemisphere. However, there is no unanimity and several Stroop studies contain a number of methodological limitations (e.g., target naming). In reality, the results that are most favorable to a semantically competent right hemisphere have come from semantic priming experiments. The right hemisphere's potential is greatest for automatic lexicosemantic processes. This observation is compatible with Hughlings-Jackson's (1915) hypothesis that the true functional lateralization of the brain for words resides in the left hemisphere's dominance for the propositional use of words since both hemispheres are capable of sustaining the automatic use of words.

It should be remembered that this review has dealt for the most part with the comprehension (in opposition to the production) of isolated written words. Furthermore, and except in a rare number of cases, most of the studies have used young college or university students as subjects. It would be important to see if similar results are obtained with other population samples. One of the factors justifying this endeavor is the role of the level of schooling in the use of cognitive strategies that might influence the manifestations of the functional lateralization of the brain. Finally, studies of individuals with intact brains, just like those of individuals with split brains, are pertinent mainly with respect to the competence or the potential of one hemisphere compared with the other. They attempt to demonstrate that one hemisphere is superior, equal, or inferior to the other with respect to a given task. Very few studies have shown that the right hemisphere's contribution is necessary in tasks where only the left hemisphere is tested. As a result, these studies provide little evidence as to the true semantic contribution of the right hemisphere when the two hemispheres are simultaneously called upon to process a word together.

Right-Brain–Damaged Studies

If the right hemisphere of the right-handed individual does contribute to the semantic processing of words under normal conditions, then partial or complete right-hemisphere destruction should affect this contribution. This issue is examined in the following paragraphs, which describe the findings from studies of lexicosemantic functioning in right-handed, right-brain–damaged individuals. The special case of crossed aphasia, however, is not discussed here. Indeed, such cases are exceptional and of little importance with respect to the usual contribution of the right hemisphere to linguistic function; they reflect instead an atypical functional organization of the brain (Joanette, 1989).

Early Studies

Jon Eisenson (1959b) was one of the first to examine the hypothesis of a linguistic deficit subsequent to a lesion of the right hemisphere in the right-handed individual. This interest stemmed from his clinical impression that "persons who had incurred brain damage of the right cerebral hemisphere were not free of linguistic impairment even though they showed no obvious aphasic involvement" (Eisenson, 1962, p. 49). To evaluate his clinical impression, Eisenson administered a series of standardized sentence completion and word-definition tasks to a number of right-brain–damaged patients (Eisenson, 1959a, 1959b).

To begin with, Eisenson (1959b) observed that right-brain–damaged patients produced just as many correct or acceptable definitions as one would expect from normal subjects. However, the definitions given were more egocentric and concrete (Eisenson, 1959a). Often words were defined in terms of the use of the objects that they represented, and even more so in terms of the limited and specialized function of each object. Eisenson (1959b) also noted that the right-brain–damaged patients performed less well on the sentence completion tasks than on the word-definition tasks. Furthermore, this difference in performance on the two types of tasks was superior to the one observed in normal subjects. Finally, depending on their age (<49 years vs. 50–69 years), right-brain–damaged subjects preferred certain types of definitions to others, even if these definitions were not the most appropriate. Eisenson (1959b) concluded that a lesion of the right hemisphere results in a loss of the most abstract attitude, similar to that which, according to Goldstein (1948), affects aphasics. Eisenson added that with respect to intellectual and linguistic functioning, "The right hemisphere is neither a silent hemisphere nor a spare to be used only in the event of failure of the other hemisphere" (1959a, p. 10). The right hemisphere is essential for conceptualizations and therefore essential for language, insofar as language is a means by which concepts are communicated.

In a continuation of his initial study, Eisenson (1960) tested more subjects, shortened certain tasks, and added a control group in order to better evaluate the difficulties of right-brain–damaged patients. He observed that these patients performed significantly less than normals when they had to provide word definitions. Indeed, right-brain–damaged subjects provided fewer correct definitions and often stated that they did not know how to define certain words. This difference between the two groups was still evident even when they were comparable in terms of age and schooling. The two groups also differed when the task was to select the word corresponding to a particular definition. Furthermore, right-brain–damaged patients obtained significantly lower scores than normals on sentence completion tasks. Once again Eisenson concluded that a right-hemisphere lesion in the right-handed individual could be accompanied by a language disorder. He added that, insofar as the tasks used allow for an evaluation of verbal "intelligence," it is likely that a decrease of intelligence is also associated with the presence of a right-hemisphere lesion.

In yet another follow-up study involving an even greater number of subjects, Eisenson (1961) obtained similar results, thereby confirming his earlier findings. Thus right-brain–damaged patients had significantly more difficulty in completing sentences, particularly when the missing word was abstract. Despite the fact that there were no statistically significant differences between right-brain–damaged patients and normal controls – matched for age and schooling – with respect to the number of definitions attempted or the number of words correctly defined, a significant difference was observed when the subjects had to recognize the defined words (Eisenson, 1961, 1962, 1973). Eisenson (1962) concluded that the right hemisphere contributes to the elaboration, production, and storage of concepts in memory. Eisenson viewed this participation of the right hemisphere in the conceptualization of the world as essential for language behavior. Eisenson (1962, 1973) added that this participation is particularly important for "higher order" linguistic components as well as for more abstract linguistic elaborations. This led him to speculate that linguistic disturbances subsequent to a lesion of the right hemisphere should not be sought in normal conversation in which the use of language does not have to be exact. According to him, the affected linguistic components are indeed more "subtle."

Eisenson (1960) also provided evidence suggesting that the lexical disturbances of the right-brain–damaged patients were due, at least in part, to factors that go beyond the linguistic processing of isolated words. On the one hand, Eisenson observed that a lesion of the right hemisphere was associated with greater difficulty in establishing an individual adjustment in response to the linguistic productions of another individual. Today, we usually refer to such behavior as having something to do with pragmatics (see Chapter 7). On the other hand, Eisenson reported that the right-brain–damaged patients had more difficulty changing their response sets in the word-definition tasks, which required the production of many different definitions. Indeed, when a definition was produced, parts of it were repeated in subsequent definitions, thereby resulting in poor performance in terms of both content and form. This finding suggests that the performance of right-brain–damaged patients is affected to some extent by functional rigidity and/or perseveration.

Eisenson was not the only investigator at the time to ascertain that right-brain–damaged patients suffered from some form of linguistic impairment. For example, Weinstein and Keller (1963; see also Weinstein, 1964) compared the object-naming performance of left-brain–damaged patients, patients having sustained widespread and deep lesions, and patients with lesions of the right hemisphere. The investigators reported that the performance of the right-brain–damaged patients was quantitatively and qualitatively different from the performance of the left-brain–damaged subjects, though it was similar to the performance of the patients with widespread and deep lesions. Furthermore, they observed that a significant number of naming errors were produced by those right-brain–damaged patients who exhibited disorientation with respect to place and time, self-identification difficulties, and hemineglect or verbal anosognosia. These errors pertained, for the most part, to objects that were related to the hospitalization or incapacities of the patients (e.g., needle, wheelchair) and attested to the personal history, past or present, of the individual. Generally speaking, the name of the target object was substituted by the name of an object that was similar to the target object in terms of its form or function. Weinstein (1964) also noted that right-brain–damaged patients used more descriptive terms, adjectives, and personal information, thereby showing their familiarity with the names of the target objects.

The errors described by Weinstein (1964), as well as by Weinstein and Keller (1963), were thought to be indicative of a change in the relationship between an individual and his environment, which itself stemmed from a disturbance of "reticulohemispheric connections." The rationale behind this interpretation is as follows. According to Weinstein and Keller, the reticular system is involved in the interactions between the individual and the environment and allows for certain analogies to be made with respect to past or anticipated events. Furthermore, a word derives its meaning from several different levels of analysis: phonetic, syntactic, semantic, and even social. Indeed, the meaning of a given word depends on the social context in which it is used, on the individual's past and anticipated experience of the speaker, on the individual's personal values, and on the cultural systems in which it is used. According to Weinstein and Keller, it follows that word-naming errors, subsequent to a lesion of the right hemisphere, are most likely to occur at the social level. These errors are related to subject–environment interactions and imply immediate situations, past experience, and anticipations. It is interesting to note the similarity between Weinstein and Keller's interpretation and various other findings. These include the more recently hypothesized disturbance of the pragmatic aspects of communication behavior in right-brain–damaged subjects, the taking into account of the connotative aspects of the meanings of words processed by the right hemisphere in the commissurotomized patient, and the egocentric and concrete nature of word definitions provided by right-brain–damaged patients as reported by Eisenson (1959a).

At about the same time, Marcie et al. (1965) conducted a study on naming performance in right-brain–damaged patients. However, they found no naming disorder that could be qualified as lexicosemantic. The naming difficulties observed in some of the patients were attributed to perseveration behavior. In another study, Newcombe, Oldfield, Ratcliff, and Windfield (1971) reported that the picture-naming performance of a group of right-brain–damaged patients was, generally speaking, similar to that of normal hospitalized individuals. Along the same lines, Oldfield (1966a, 1966b) looked at the number of brain-damaged patients who provided the correct name of an object as a function of the word's frequency of use. Results showed that there was no significant difference between right-brain–damaged patients and nonaphasic, left-brain–damaged patients or normal hospitalized patients. However, both Oldfield (1966a, 1966b) and Newcombe et al. (1965) pointed to the longer amount of time required by right-brain–damaged patients to name the pictures. In fact, Oldfield reported that right-brain–damaged patients were significantly slower than nonaphasic, left-brain–damaged subjects on naming tasks. Their slowness was not due to lesion size since the difference was still significant even when this factor was controlled for. Furthermore, the slowness of aphasics was more important for low-frequency than for high-frequency words, whereas the slowness of right-brain–damaged patients was independent of the frequency of word usage. Oldfield (1966a) attributed this slowness to a change in visual perception resulting in a blockage of normal recognition of the pictures to be named. More precisely, the right-brain–damaged subjects' slowness was considered more indicative of an alteration of the visual field than of a disorder of object identification per se.

According to the performances reported by Archibald and Wepman (1968), not all right-brain–damaged patients are language disordered. This observation emerges from the application of the language modalities test for aphasia constructed by Wepman and Jones (1961) in order to evaluate some components of language such as repetition, writing, reading, and naming. Of the 22 right-brain–damaged patients tested by these investigators, only 8 showed any signs of a language disorder. Interestingly, the performance of three of these eight patients was within normal limits. The errors of a fourth patient were the result of a left visual field impairment, while yet another patient was classified as truly aphasic due to the semantic nature of his difficulties on a storytelling task. Archibald and Wepman noted that their language-disordered, right-brain–damaged patients almost never made semantic errors, nor were their errors similar to those described by Weinstein (1964).

In light of the nonlinguistic performance of right-brain–damaged patients (e.g., Raven Progressive Matrices, Sorting Test, visuoconstructive test, maze test), Archibald and Wepman suggested that the presence of a language disorder in right-brain–damaged patients was the result of their generally inferior level of intelligence. According to the authors, individuals with low intelligence scores pay less attention to tasks. It is therefore plausible that certain patients obtain lower scores on linguistic and nonlinguistic tasks as a result of their poorer attention skills. Therefore, it might be that the language disorders observed in certain right-brain–damaged patients are more the result of a global deterioration of intelligence and are not, a priori, of a lexicosemantic nature. In a critical review of Eisenson's (1962) work, Archibald and Wepman noted that no details were given concerning the intellectual level of the patients involved. This factor could have had a significant influence on the results of his study. In fact, the same could be said of Weinstein's (1964) study. Finally, given the absence of any pertinent information, Archibald and Wepman did not exclude the possibility that lesion site is a determining factor in helping to explain why not all right-brain–damaged patients are affected.

In summary, the results of studies conducted during the sixties, in order to systematically verify the presence of lexicosemantic disorders in right-brain–damaged subjects, show a lack of consensus. Not only are such disorders not always present, but when they are, not all right-brain–damaged patients are affected. Furthermore, such disorders are sometimes considered as secondary to other nonspecific linguistic or nonsemantic disorders. On the basis of these findings, a twofold controversy emerges. First, can a lesion of the right hemisphere of a right-handed individual affect the processing of word meaning? Second, if the functional deficit subsequent to a lesion of the right hemisphere does indeed affect the processing of words, is the resulting impairment one that is purely lexicosemantic in nature? Recent studies have attempted to answer these questions.

Word Comprehension

Ruth Lesser (1974) was one of the first to have conducted a systematic investigation of word-comprehension capacities in right-brain–damaged patients. The main objective of this study was to evaluate the clinical usefulness of the English versions of three tests originally developed for native Italian-speaking subjects. These tests were designed to evaluate auditory comprehension capacities at the phonological, semantic, and syntactic levels. Lesser's patients were also tested on a battery of more conventional tests designed to assess their syntactic (Token test), lexical (English Picture Vocabulary Test—EPVT), and intellectual capacities (Raven's Progressive Matrices). The "new" semantic test required the subjects to point to the picture of a word after it had been spoken by the examiner. A multiple-choice array of four pictures was presented on each trial. Each array contained the picture of the target word as well as three other pictures representing objects that were frequently associated with the target word. For example, if the target word was paper, the subjects had to choose between the pictures of a pencil, a pen, several sheets of paper, and a person writing. The four pictures were arranged in a square. The difference between this task and the EPVT is that in the latter (1) the pictures are arranged in a single row, (2) the distractors are not systematically associated with the target, and (3) the target words are of increasing difficulty. Results for the "new" semantic task showed that right-brain–damaged patients performed significantly worse than normals but produced scores comparable to those of aphasics. Furthermore, the right-brain–damaged patients obtained scores comparable to normal controls on the phonological and syntactic tests, but they performed significantly worse than the controls on the EPVT and Progressive Matrices, where their performance was comparable to that of aphasics. One can therefore conclude, on the basis of these findings, that the two lexicosemantic tasks used by Lesser reveal a deficit in right-brain–damaged patients.

The relatively poor performances of Lesser's (1974) right-brain–damaged patients could not be attributed to any visual impairment. On the one hand, Lesser reported that none of the right-brain–damaged patients showed any clinical evidence of visual agnosia. On the other hand, the right-brain–damaged patients and the normal controls obtained comparable scores on the phonological task. The presentation format of this task was similar to that of the semantic task previously described. In an attempt to uncover some of the factors involved,

Lesser discussed the eventual role of bilateral lesions of the brain. However, there was not enough clinical evidence to support or refute this explanation. Yet it remains plausible if one assumes, in light of the fact that the right-brain–damaged patients showed no phonological or syntactic disorders, that the lexicosemantic aspects of language are more sensitive to the presence of a lesion of the left hemisphere. Lesser also examined the hypothesis that the presence of a more global deterioration of intelligence may have affected the performance of the right-brain–damaged patients. In fact, the rather low scores on the Progressive Matrices, as well as the correlation between scores on the Raven test and scores on the three adapted subtests, for all subjects combined, are compatible with such an interpretation. Lesser emphasizes the fact that the worst performed tasks involved series of pictures that the subjects had to systematically examine and consider before giving an answer. Thus it appears that the right-brain–damaged patients were more inclined to accept an approximation of the answer than to search for the exact answer. Finally, another factor that may have played a role is the manual preference of the subjects: Lesser did not provide any details concerning the proportion of true right-handers within the group of right-brain–damaged patients.

Whatever the reasons for the impaired performance of her right-brain–damaged patients, Lesser's (1974) most interesting finding was the preservation of syntactic discrimination contrasting with semantic impairment. In light of the specific lexicosemantic nature of the impairment observed in these patients, Lesser hypothesized that the right hemisphere contributes, in one way or another, to the comprehension of words, and perhaps more specifically to word selection when presented among semantically associated words. According to Lesser, the eventuality that the aforementioned impairments reflected a more remote interference of left-hemisphere functions (diaschisis) was not a plausible explanation, unless one postulates that the same disturbed processes would also be sustained by the right hemisphere.

In an attempt to evaluate this hypothesis, Gainotti, Caltagirone, and Miceli (1979) sought to determine the extent to which the presence of visual neglect or general mental deterioration could account for the semantic errors made by right-brain–damaged patients on an auditory verbal comprehension test. All of the right-brain–damaged patients had unilateral cerebral lesions subsequent to vascular accidents (N = 72) or tumors (N = 38). The controls (N = 94) included both psychiatric patients and patients with lesions of the peripheral nervous system. All of the subjects were right-handed and, as in the Lesser (1974) study, the two groups were similar with respect to age and schooling. The auditory verbal comprehension test was similar to the task used by Lesser (1974): a concrete noun was orally presented by the examiner and the subject had to point to the corresponding picture contained in an array of six pictures arranged in two columns of three. Each of the arrays contained a semantically related distractor, a phonemically related distractor, and three unrelated foils.

To begin with, Gainotti et al. (1979; see Gainotti et al., 1983) found no significant difference between right-brain–damaged patients and controls with respect to the number of phonemic errors. However, a closer examination of semantic errors revealed that the right-brain–damaged patients chose the semantically related distractor more often than the controls. Based on their performance on a series of verbal and nonverbal memory and intelligence tests, the right-brain–damaged patients were classified as either "mentally deteriorated" or "non–mentally deteriorated." Gainotti et al. (1979) observed that the semantically related distractor was chosen significantly more often by the patients designated as mentally deteriorated than by the controls, whereas the non–mentally deteriorated patients chose the semantically related distractor only slightly—although significantly—more often than the controls. Finally, Gainotti et al. (1979) examined in detail the trials in which the semantically related distractor was chosen when it appeared, along with the target word, in the left-hand column in the array. According to Gainotti et al. (1979), in the event of a bias introduced by the presence of left unilateral neglect, the presence of a semantic error will be ascertained if the semantically related distractor is located in the same column as the target word. This analysis revealed a much greater decrease in the mean number of semantic errors made by the right-brain–damaged patients than by the control subjects. In fact, there was no significant difference between right-brain–damaged patients and control subjects with respect to the number of subjects who made at least one semantic error of this type under these conditions.

Notwithstanding their finding that the right-brain–damaged patients made significantly more semantic errors than the controls on an auditory verbal comprehension test, Gainotti et al. (1979) concluded that there was a relationship between mental deterioration and the number of semantic errors made. They also argued that there was no semantic disorder in the right-brain–damaged patients when the analysis took into account the eventual presence of left unilateral neglect. In addition, Gainotti et al. drew attention to the fragile nature of lexicosemantic organization from the neuropsychological point of view. They assumed that following right-hemisphere damage, semantic confusions occurred only in the event that the correct response was not immediately available and/or in the presence of a more global mental disorder. Paradoxically, Gainotti et al. simultaneously accepted as well as rejected the possibility that the right hemisphere specifically contributes to the semantic processing of words. However, in a subsequent summary of this study, Gainotti et al. (1983) claimed that visual hemineglect and/or a more general cognitive disorder were only partly responsible for the semantic errors made by right-brain–damaged patients: such errors were not to be entirely attributed to these two factors.

In a second study, Gainotti, Caltagirone, Miceli, and Masullo (1981) attempted to dissociate the phonemic and semantic components of words by using two different tasks. Once again, they sought to determine if the errors made by their subjects were the result of a linguistic disorder. Thus they took into account possible sources of interference such as left visual hemineglect or global mental deterioration. They also sought to determine if these errors were specific to auditory comprehension or if they also occurred in reading comprehension. Gainotti et al. therefore administered a phonemic discrimination task and two semantic comprehension tasks to a group of right-brain–damaged patients and a group of normal controls. Each of the semantic tasks consisted of associating a spoken or read target word with its corresponding picture. This picture was part of a triad presented in a single vertical column in a way that would minimize any possible effects of visual hemineglect. The distractors were semantically related to the target (e.g., target: fork; distractors: knife and spoon). The right-brain–damaged patients made significantly more semantic errors than the normal controls but showed no impairments on the phonemic task. However, Gainotti et al. reported that on each of the three tasks the right-brain–damaged patients with some degree of mental deterioration made significantly more errors than both the right-brain–damaged patients who were not mentally deteriorated and the normal controls. Conversely, right-brain–damaged patients with no mental deterioration made significantly more errors than controls on the auditory verbal comprehension task. They also made more errors than the controls on the reading task, although this was only a nonsignificant trend. There was no significant difference between auditory and reading comprehension errors, although the mean number of semantic errors was somewhat higher for orally presented words than for words that were read.

Gainotti et al. (1981) suggested that the degree of mental deterioration observed in their right-brain–damaged patients may have had an influence on the number of semantic errors made. These researchers do admit, however, that the lexicosemantic disorder observed in their patients may have been directly related, at least in part, to the presence of a right-hemisphere lesion rather than to global mental deterioration or to the presence of unilateral visual neglect. The findings of this study also point to the presence of a specific lexicosemantic deficit in the right-brain–damaged patients since no significant differences were observed between these patients, in particular the nondeteriorated patients, and normal controls on the phoneme-discrimination task. However, considering the fact that the semantic comprehension tasks required the use of picture materials, Gainotti and his colleagues do not exclude the possibility of subtle visuoperceptual disorders in order to explain the performance of their right-brain–damaged patients.

Finally, Gainotti et al. (1981) hypothesized that complex linguistic tasks, such as the semantic task used in their study, necessitate the participation of the two cerebral hemispheres. Thus the right hemisphere of the right-handed subject contributes to semantic functioning, at least in certain tasks. However, Gainotti et al. do not entirely exclude another explanation: the semantic errors of right-brain–damaged patients are due to higher order cognitive impairments rather than to a purely linguistic or lexical deficit. Such a hypothesis is based on the relationship observed between the semantic deficit and the degree of mental deterioration. Mentally deteriorated right-brain–damaged patients make more semantic errors than nondeteriorated patients who make just as many errors as normals. Gainotti et al. concluded that the mild semantic deficit of the nondeteriorated right-brain–damaged patients could be in reality the result of mild mental deterioration that goes somewhat unnoticed in the clinical examination. Under these conditions, a deficit subsequent to a lesion of the right hemisphere does not appear to be specific to the semantic processing of words.

According to Bishop and Byng (1984), the comprehension tasks used by Lesser (1974) and Gainotti et al. (1981) failed to provide a true indication of the semantic nature of the errors made. To begin with, all of the distractors were semantically related to the target word. As a result, the errors could only be "semantic," regardless of the actual underlying impairment (e.g., phonological, unilateral neglect, or lack of motivation). Second, Bishop and Byng believed that some of the semantic distractors previously used had a number of visual characteristics in common with the target object. For example, the distractors used by Gainotti et al. for the target "fork" were "spoon" and "knife." All three objects have a long, narrow shape and are usually of the same color. In this particular instance, the "semantic errors" may have been visual rather than linguistic in origin. In their critical review, Bishop and Byng did not take into account the results of Lesser's phonological task, which seem to reject the hypothesis of a phonological or visual disorder to explain the difficulties observed in her semantic task. Nor did Bishop and Byng take into consideration the Gainotti et al. (1979) study in which phonological, semantic, and unrelated errors were possible. Nonetheless, Bishop and Byng developed a comprehension task (auditory and reading) that was supposedly effective in distinguishing visual problems from semantic deficits.

The task used by Bishop and Byng (1984) involved the recognition of a target object presented within an array of either four or eight pictures. Four testing conditions were used, depending on the relationship between the target and one of the distractors, while the other distractors (i.e., three or seven) were unrelated to the target. The relationship between the target and the controlled distractor was visual (e.g., pipe–hatchet), semantic (e.g., sister–church), visual-semantic (e.g., cat–fox), or absent.

Bishop and Byng (1984) found few errors on their task. Those of the right-brain–damaged patients essentially involved distractors having a visual or visual-semantic rather than a semantic relationship with the target. The three groups of subjects that were involved in this study (right-brain–damaged patients; left-brain–damaged patients, including seven nonaphasics, seven recovering aphasics, and three aphasics; and normal hospitalized subjects) produced a statistically similar number of errors. Bishop and Byng suggested that subjects with a unilateral lesion and those showing real semantic disorders are rare and that a lesion of the right hemisphere does not result in a disorder that can be qualified as semantic when nonlinguistic (e.g., perceptual) factors are controlled for. A second explanation put forward by Bishop and Byng was that their semantic task was not difficult enough to reveal a semantic disorder. This explanation was rejected for two reasons. First, Bishop and Byng reported that response latencies failed also to show that the semantic distractors were particularly difficult. Second, some left-hemisphere–damaged patients produced evidenced of a semantic comprehension deficit. However, the first argument does not really rule out the possibility that the task involving the semantic distractor was too easy. As for the second argument, it is possible that the Bishop and Byng task was sensitive enough to the presence of a semantic deficit associated with aphasia or visual agnosia, but was not sensitive enough to show semantic deficits in right-brain–damaged patients. Furthermore, Code (1987) drew attention to another point to account for the absence of any semantic deficit in the patients tested by Bishop and Byng. Code emphasized the fact that these patients had suffered hemispheric damage subsequent to a missile wound incurred when they were young, 35 to 40 years before this study was conducted. Compared to the subjects of the other studies reviewed here, it is highly probable, still according to Code, that these patients had more restricted cerebral damage and that their recovery was better given their age at the time of the accident and the amount of time postonset.

If right-brain–damaged patients do in fact have a lexicosemantic deficit, Gainotti et al. (1983) proposed that the manifestation of this disorder should not be restricted to word comprehension but should also include word-production abilities. In an attempt to evaluate this hypothesis, the authors administered a picture-naming task to a group of right-brain–damaged patients. The errors were classified into visual (e.g., "ball" for "apple"), semantic (e.g., "pear" for "apple"), or visuosemantic (e.g., "peach" for "apple"). Results showed that mentally deteriorated right-brain–damaged patients made the most errors whether or not visual and semantic errors were considered separately. Nondeteriorated right-brain–damaged patients produced mainly semantic and visuosemantic errors. When the number of subjects who made at least one error of any type was considered, Gainotti et al. reported a significant difference between the two groups for visuosemantic errors but not for visual or semantic errors. The difference between the two groups for semantic errors was very close to the level of statistical significance. Once again, Gainotti et al. concluded that the general mental deterioration of certain right-brain–damaged patients affects their performance. Furthermore, the comprehension and/or production errors made by right-handed, right-brain–damaged subjects cannot be entirely attributed to a visuoperceptual disorder since errors also attest to the presence of some genuine lexicosemantic disorder.

The lack of methodological details concerning their classification of naming errors casts some doubt on the procedure employed by Gainotti et al. (1983). Indeed, even if judges were in agreement as to the classification of a given error, their judgments would remain arbitrary considering the nature of the categories used. Thus in the example given earlier, where the stimulus is "apple," the response "pear" is classified as a semantic error, whereas "peach" is classified as a visuosemantic error. However, there is nothing to demonstrate that this classification is the most appropriate. For example, naming an "apple" a "pear" or a "peach" could be classified as either a semantic error or a visual confusion.

The issue of the linguistic specificity of language disorders in right-brain–damaged patients was also addressed in a study conducted by Coughlan and Warrington (1978). The subjects were administered a large battery of tests that included Raven's Progressive Matrices and the Wechsler Adult Intelligence Scale (WAIS) vocabulary subtest, which provided an estimate of each subject's IQ. The other tasks included a repetition task involving monosyllabic words used to assess the subject's articulation abilities; a phoneme-discrimination task where the subjects had to indicate if two monosyllabic words were identical; and a modified version of the Token Test in which the subjects had to understand and execute increasingly complex commands. A second series of tasks dealt more specifically with comprehension and word-access abilities. These tasks included a modified version of the Peabody Picture Vocabulary Test in which a spoken word has to be associated with its corresponding picture presented in a multiple-choice array; the Auditory Choice Vocabulary Test in which subjects have to specify which of two words is semantically related to a target word; a visual object-naming task; and a naming task involving orally described objects.

Generally speaking, the results of Coughlan and Warrington (1978) showed that the performance of right-brain–damaged patients was similar to that of normal controls on the repetition, phoneme-discrimination, and object-naming tasks, as well as on the WAIS vocabulary subtest. However, the two groups differed significantly on the Progressive Matrices, the Token Test, and the Peabody Test. In addition, right-brain–damaged patients tended to score lower than controls on the Auditory Choice Vocabulary as well as on the naming task involving oral descriptions. According to Coughlan and Warrington, several of the tasks that were poorly performed by the right-brain–damaged patients involved the comprehension of word meanings. However, it is unlikely that the origin of their difficulties occurred at this level. Indeed, Coughlan and Warrington observed that the right-brain–damaged patients had few problems with the WAIS vocabulary subtest, which is known to be sensitive to left-hemisphere lesions. They add that the Token Test, one of two linguistic tasks that were particularly difficult for the right-brain–damaged patients, requires very little competence in terms of word comprehension. Furthermore, it is unlikely that the deficits observed were the result of bihemispheric dysfunctioning. Indeed, right-brain–damaged patients were able to complete the object-naming task, which also seems to be particularly sensitive to left-hemisphere lesions. In reality, the most plausible explanation, at least according to Coughlan and Warrington, lies at the level of the organization of adequate strategies in response to the demands of the more difficult tasks. Coughlan and Warrington were unable to define the exact nature of these strategies. However, they suggested that the problem of right-brain–damaged subjects occurs at a higher level of cognitive functioning and does not correspond to a specific disorder of linguistic functioning per se.

In summary, previous studies have failed to provide any solid evidence to the effect that a lesion of the right hemisphere results in a disorder that one can describe as being exclusively lexicosemantic. The problem resides therefore in the linguistic specificity of the disorders underlying the difficulties of right-brain–damaged patients to process word meaning. The possibility of a lexicosemantic contribution by the right hemisphere is purely hypothetical, although there is some indication that the right hemisphere of the right-handed individual is essential for the full semantic processing of words.

Semantic Judgments

The lexicosemantic abilities of right-handed, right-brain–damaged individuals can also be examined using semantic judgment tasks. These tasks require that a decision be made concerning the existence of a semantic link between at least two words or pictures. This particular type of task is referred to as metalinguistic, since it calls for a judgment that is based on an individual's linguistic knowledge. Semantic judgment tasks differ from more conventional situations of verbal communication. As a result, one must be careful when using the findings from semantic judgment studies to draw conclusions about normal communication abilities.

Antonymy and Synonymy

The normal adult has a general understanding of antonymy (Gardner, Silverman, Wapner, & Zurif, 1978). For example, at the linguistic level, he understands that *A* is opposed to non-*A* or that cat is opposed to mouse. At the nonlinguistic, or conceptual, level, he understands that depth is opposed to height, as is noise to silence, and that an arrow pointing up is opposed to an arrow pointing down. On the basis of this assumption, Gardner et al. conducted three experiments. The first concerned the production of canonical (i.e., clearly designated and accepted) and noncanonical (i.e., plausible) linguistic opposites. Subjects were instructed to provide the best opposite for each word that was presented. The stimuli included 20 words arranged in canonical pairs with well-established opposites (e.g., prince–peasant). The stimuli also included 14 noncanonical lexical items that normally had no direct opposites (e.g., ocean–desert; salt–pepper). Results showed that right-brain–damaged subjects scored higher than left-brain–damaged subjects and performed just as well as normal controls, at least for canonical opposites. From a qualitative viewpoint, Gardner et al. considered all the possible types of incorrect productions and found that the right-brain–damaged subjects had a tendency to produce more synonyms and semantic associations than the left-brain–damaged subjects. Finally, the distribution of the types of incorrect productions made by controls was more similar to that of the aphasic than of the right-brain–damaged subjects.

In a second experiment, subjects had to choose the correct opposite from a multiple-choice array. Three types of stimuli were used: pictorial representations (i.e., cartoonlike pictures), linguistic stimuli (i.e., words), and abstract designs (i.e., no linguistic referent). The response choices for each stimulus included its direct opposite (i.e., the correct response), a synonym, a partial opposite corresponding to an item halfway between the two opposing poles, and an item that was totally unrelated to the stimulus. If, for example, one of the stimuli presented in the pictorial representation condition was the drawing of a short man, then the four choices would include the pictures of (1) a tall man, (2) an even shorter man, (3) a man of average height, and (4) a fat man. Gardner et al. (1978) report that in the condition involving linguistic stimuli, the performance of right-brain–damaged subjects was similar to the performance of anterior aphasics and superior to that of posterior left-brain–damaged subjects. However, in the two conditions involving pictorial stimuli, right- and left-brain–damaged subjects performed in much the same way. Finally, an inspection of the errors produced showed that Broca's aphasics and normal subjects chose partial opposites more frequently, whereas Wernicke's aphasics and right-brain–damaged subjects chose synonyms more often. According to Gardner et al., these two groups of subjects demonstrated a decreased awareness of the notion of antonymy.

The third experiment conducted by Gardner et al. (1978) involved pairs of words that were presented either orally or visually. Following each presentation, subjects had to indicate nonverbally if the two words were antonyms. The word pairs consisted of (1) optimal opposites (e.g., difficult–easy), (2) unrelated items (e.g., difficult–dark), or (3) semantically related items that included synonyms (e.g., difficult–hard), less potent antonyms (e.g., death–beginning), or partial opposites (e.g., truce–war). No significant difference was found between the performance of right-brain–damaged subjects and that of aphasics with anterior and/or posterior lesions. However, in the case of the semantically related word pairs, right-brain–damaged subjects were more inclined than left-brain–damaged subjects to indicate that the two words were antonyms, and left-brain–damaged subjects were more likely to judge unrelated items as opposites. On the basis of these findings, Gardner et al. concluded that left-brain–damaged subjects were more similar to normal controls than to right-brain–damaged subjects. Thus in each of the three experiments reported by these investigators, right-brain–damaged subjects had a special status.

According to Gardner et al. (1978), right-brain–damaged subjects generally perform at a high level except when nonverbal materials are involved. However, the errors produced by these subjects suggest an overall weakening of the notion of antonymy. In reality, right-brain–damaged subjects obtain better scores when the opposites are overlearned or canonical, that is to say when the production or the selection of an opposite is quasi-automatic. Conversely, as soon as the opportunity to commit errors increases, as is the case for the production of noncanonical opposites or for distracting pairs in the antonymy-judgment task, right-brain–damaged subjects appear to be more attracted by nonopposing items. According to Gardner et al., this difficulty in rejecting plausible combinations can be attributed to the absence of a solid anchor at one of the opposing poles of the concept to be processed. Gardner et al. admit that the origin of the poor sensitivity of right-brain–damaged subjects to antonymic relationships has yet to be established. Among the various hypotheses entertained by these researchers is a decrease in the conceptual capacities of right-brain–damaged patients. Gardner et al. also hypothesized that this decrease in sensitivity could be the manifestation of a weakness of the spatial imagery mechanism, which usually serves as a basis for the opposition of concepts.

In contrast to the notion of antonymy, which is based on words that have opposite meanings, synonymy refers to words that have the same or nearly the same meaning. A recent experiment conducted by Goulet and Joanette (1988) focused on this type of semantic relation in order to address the issue of the contribution of the right hemisphere to the semantic processing of abstract words.

Indeed, many studies of the lexicosemantic capacities of brain-damaged patients had shown that right-brain–damaged subjects are much less effective than normals when dealing with concrete words. However, these difficulties could have been associated with a higher level cognitive disorder rather than a specific linguistic impairment. Assuming that abstract words are not "represented" or accessible by the right hemisphere (cf. tachistoscopic hemifield viewing experiments with normals), Gainotti et al. (1981) proposed to test this hypothesis by examining the ability of right-brain–damaged subjects to process abstract words. If, according to the authors, the difficulty of right-brain–damaged subjects lies at the linguistic level, then these patients should be able to process abstract but not concrete words. If, however, their problem lies at a higher cognitive level, then they should evidence difficulties with both types of tasks. Thus, in their study, Goulet and Joanette (1988) used abstract noun triads, each containing two synonyms or quasi-synonyms and one distractor that was associated with at least one of the target words on the basis of a syntagmatic relationship (e.g., story–miracle–anecdote). Following the presentation of each triad, the subject had to indicate the two words that "go together best." The results revealed no statistically significant difference between right-brain–damaged subjects and neurologically intact controls. Goulet and Joanette admitted that their results should be considered cautiously. On the one hand, the study involved few right-brain–damaged subjects (N = 14). On the other hand, it might be that the task was too easy. Nonetheless, the results of this study appear to be consistent with the hypothesis that the integrity of the right hemisphere is not essential for all forms of abstract word processing. The results are also more compatible with the presence of a lexicosemantic deficit rather than a more global cognitive deficit in right-brain–damaged patients. Following a suggestion made by Brownell, Bihrle, Potter, and Gardner (1985), Goulet and Joanette argue that right-brain–damaged patients perform well when they have to rely solely on their representations of highly systematic, dictionary-type lexicosemantic knowledge.

Connotation and Metaphor

Gardner and Denes (1973) developed a task to examine the sensitivity of aphasic and right-brain–damaged subjects to the connotative meaning of words. Subjects were instructed to select which of two line drawings best illustrated the connotative attributes of an orally presented word (e.g., woman). The stimulus words were either concrete or abstract nouns and adjectives. During the course of a preliminary study, Gardner and Denes had searched for natural associations, in terms of common connotations, between stimulus words and line drawings. In all, 31 words and 14 pairs of drawings were selected. The two items composing each pair of line drawings were differentiated by a visual characteristic that was easily perceivable. Given the low number of right-brain–damaged subjects (N = 6), Gardner and Denes were unable to conduct a quantitative analysis and resorted instead to a qualitative analysis of their subjects' responses.

To begin with, Gardner and Denes (1973) emphasized the fact that two right-brain–damaged subjects refused to perform the task altogether and that a third took so much time that he was unable to finish. Even more interesting was the fact that all of the right-brain–damaged subjects were more or less opposed to performing the task, which, according to the authors, seldom occurs in aphasics. For example, one subject responded to the stimulus word "woman" in the following way: "Neither of these lines is a woman. I don't see any woman there. What are you trying to pull?" After having reassured each subject that there was indeed no woman in the drawing and that the experimenter wanted only to know the subjects' impressions concerning the drawing that was most reminiscent of a woman, three of the subjects eventually completed the task. These subjects obtained performances of 66%, 71%, and 74%, all of which were inferior to the performance levels of aphasics with anterior lesions. Gardner and Denes pointed out that although these right-brain–damaged subjects performed poorly, they were not known to have any linguistic disorders.

In discussing their observations, Gardner and Denes (1973) draw attention to some known characteristics of right-brain–damaged subjects: changes in emotional state, more rigid and concrete behavior, difficulties in metaphorical thinking to the extent that they have difficulty understanding the figural meaning and/or the less literal meaning of words. They also discuss the possibility of a decrease in their subjects' intellectual functioning or the use of a defensive strategy in the wake of their illness. Finally, Gardner and Denes suggest that the left hemisphere is more involved in the dichotomous aspects of comprehension, such as those involved in denotation (a given element denotes or does not denote this element or that), whereas the right hemisphere plays a greater role with respect to the perception of the details and the characteristic features of the connotative aspect of words. Gardner and Denes did not consider the possibility that a visuoperceptual disorder may have been at the origin of the right-brain–damaged subjects' behavior. The presence of a visuoperceptual disorder could have helped explain the defensive attitude of these subjects. It could even have helped explain their difficulty in forming a correct link between a word and a drawing which requires visual analysis, and this all the more so if the drawing has no meaning to begin with. No information was provided to this effect. Thus the true origin of the difficulties of right-brain–damaged subjects is still unknown.

Following up on Gardner and Denes's (1973) experiment, Brownell and colleagues (Brownell, 1988; Brownell, Potter, Michelow, 1984) underlined the fact that certain words can be grouped together on the basis of their denotative or connotative meanings. For example, the word "warm" can be associated with the word "cold" since both refer to temperature (denotation). However, the word "warm" can also be associated with the word "loving," since both connote a positive valence and refer connotatively to affective traits of human beings.

In an attempt to determine if connotation and denotation are indeed dissociable, which would attest to the existence of different mechanisms or lexical representations, Brownell and colleagues (Brownell, 1988; Brownell et al., 1984) presented triads of words to right- and left-brain–damaged (i.e., aphasics with pre- and postrolandic lesions) subjects, as well as to a group of normal controls. Upon the presentation of each triad, subjects had to indicate the two words that were most closely related. For example, in the triad "loving–cold–foolish," the words "loving" and "foolish" denote personality traits, whereas the terms "cold" and "foolish" could be grouped together on the basis of their negative connotations. Thus to adequately perform this task, subjects had to ignore one meaning of the reference word and focus on the other. The results revealed a double dissociation. More specifically, the right-brain–damaged subjects preferred the denotative component, whereas the left-brain–damaged subjects preferred the connotative component of the reference words. Normal subjects demonstrated a more intermediate attitude, taking into account both the denotative and connotative aspects of the words used. According to Brownell et al. (1984), these results clearly show that one cannot affirm that right-brain–damaged subjects are entirely insensitive to the denotative or connotative components of words. Nevertheless, the performance of the brain-damaged subjects suggests a certain independence between the connotative and denotative aspects of word meaning, at least from a neuropsychological point of view. According to Brownell et al. (1984), such a dissociation might imply that word meaning in the normal individual results from the unification of dissociable lexical structures sustained by either of the cerebral hemispheres. It is therefore obvious to these authors that the right hemisphere of the right-handed individual is actively involved in the cognitive, lexical, conceptual, or interpretive activities that allow access to the connotative meaning of words.

In a second study, Brownell and colleagues (Brownell, 1988; Brownell, Bihrle, Potter, Gardner, 1985; Brownell, Simpson, Bihrle, Potter, Gardner, submitted) attempted to determine whether the lexical problems of right-brain–damaged subjects are specific to the nondenotative or nonliteral meaning of words or are just the result of a difficulty in processing polysemous words. In this study, two types of words were used, each having at least two different meanings. The first type of words were adjectives (e.g., warm: heat, fire) which could have a metaphoric meaning (e.g., warm: loving). The second type were nouns which had two nonmetaphoric meanings. One meaning was frequent (e.g., pupil = structure of the eye) and the other less frequent (e.g., pupil = student). Triads were constructed with each of the target words. In addition to the target polysemous word (e.g., warm, pupil), each triad contained a nonsynonymous word associated with the most frequent meaning of the target word (e.g., blanket, eye) and a word that was quasi-synonymous with the second meaning of the target word (e.g., affectionate, student). The subject's task was to indicate the two words that were most similar in meaning. According to the instructions given to the subjects, a correct answer is contingent on the association of the target word with the word that is associated with its second meaning (e.g., warm–affectionate, pupil–student). Optimal performance on this task requires that subjects resist being attracted by the word associated with the more frequent meaning and that they consider the less frequent meaning.

Non–brain-damaged control subjects produced very few errors on this task (i.e., fewer than 2 errors over 16 trials). Of the 18 right-brain–damaged subjects, 3 also obtained near-perfect scores, thereby suggesting that a right-hemisphere lesion is not inevitably associated with impaired lexicosemantic processing. These three subjects were therefore eliminated from any further analysis. Interestingly, the remaining right-brain–damaged subjects, as well as left-brain–damaged aphasics with pre- and postrolandic lesions, obtained lower scores than the controls in both experimental conditions (i.e., metaphorical adjectives and nonmetaphorical nouns). Furthermore, Brownell and colleagues (Brownell, 1988; Brownell et al., submitted) reported a statistically significant interaction between the two experimental conditions and the presence of right or left brain damage. More specifically, left-brain–damaged patients obtained similar scores whether the target words had a metaphorical second meaning or not, whereas right-brain–damaged patients obtained higher scores when the target words had no metaphorical second meaning. The results also show that the right-brain–damaged patients perform more poorly than the left-brain–damaged patients on the "metaphorical" task, whereas it is just the opposite with respect to the "nonmetaphorical" task.

According to Jones (1983), a crossover interaction, such as the one observed in Brownell et al.'s (Brownell, 1988; Brownell et al., submitted) second study, could reflect the presence of a real double dissociation. Thus, in line with Jones's interpretation of the double-dissociation paradigm, one could infer that lesions of either the right or left hemisphere affect different functions, each of which has a different role in performing metaphoric and nonmetaphoric tasks, and the fact that right- and left-brain–damaged subjects did not score at normal levels on metaphorical and nonmetaphorical tasks respectively could be attributed to a nonspecific lesion effect. Therefore, right-brain–damaged subjects seem to have a general disorder affecting their appreciation of the alternative meanings of polysemous words (Brownell et al., submitted). In addition, they exhibit an extra difficulty in their appreciation of metaphoric meanings, which must be distinguished from the hypothesis of a nonspecific tendency to ignore any alternative interpretation in word processing (Brownell et al., 1985). Like Chiarello et al. (1987), Brownell (1988) suggests that when processing a word, the role of the intact right hemisphere seems to involve maintaining the alternative meanings of ambiguous words while the left hemisphere's role is to gradually focus on only one meaning. In addition, the right hemisphere might have a specific role to play with respect to metaphoric processing. However, the exact nature of this role has not yet been determined (Brownell, 1988).

Some of the other findings reported by Brownell (1988) and Brownell et al. (submitted) concern the role of factors such as the emotional component of the stimulus words and the frontal versus nonfrontal nature of the lesion. To begin with, the degree of semantic similarity between the literal and metaphoric meanings of the adjectives used in this study did not correlate significantly with the performance of the right-brain–damaged subjects, whereas the aphasics performed better when the target's alternative meanings were closely related. This suggests that the two patient groups processed the metaphoric triads differently. Second, the difficulty of the right-brain–damaged subjects to process metaphoric adjectives was not influenced by the emotional value of the stimulus words (i.e., more versus less emotional). Thus it appears that a purely affective factor cannot explain the poor performance of the right-brain–damaged subjects (Brownell, 1988). Finally, Brownell et al. (submitted) failed to find any evidence suggesting a special role for either the right or left frontal lobe with respect to the processing of the metaphoric meaning of adjectives.

Naturalness and Cohyponymy

Wilkins and Moscovitch (1978) examined the possible disturbance of semantic memory subsequent to an anterior temporal lobectomy in either hemisphere. Their subjects were administered two tasks in which drawings or object names had to be classified depending on whether they were bigger or smaller than a chair (e.g., lion, eye) or whether they were natural or man-made (e.g., cactus, cigarette). According to Wilkins and Moscovitch, the size judgments require an analogic, nonverbal representation system that is essentially under the control of the right hemisphere. As for the second type of judgment, it cannot rely solely on perceptual characteristics and requires, still according to the authors, reference to a verbal representation system that is under the control of the left hemisphere. The left-brain–damaged subjects in this experiment showed a selective disturbance of object classification depending on whether the objects were man-made. This finding was independent of the modality in which the stimuli were presented (i.e., words or pictures). The right-brain–damaged subjects showed no significant impairment in comparison to normal controls. To explain their findings, Wilkins and Moscovitch suggested that damage to the anterior portion of the left temporal lobe resulted in the disturbance of semantic systems involving verbal or lexical representations, whereas those systems involving analogic or visual representations remained intact. These investigators add that if a semantic disorder did exist in right-brain–damaged subjects, it would not be of the same nature as that observed in left-brain–damaged subjects, since the former subjects were able to adequately perform both tasks. Without providing any evidence as to the existence of a certain right-hemisphere–based semantic potential, Wilkins and Moscovitch suggest that if the existence of this potential were to be acknowledged, there would be qualitative differences with respect to the semantic access and/or organization sustained by each hemisphere.

In contrast to the results reported by Wilkins and Moscovitch (1978), Chiarello and Church (1986) reported evidence of a lexicosemantic deficit in right-handed, right-brain–damaged subjects. Their study consisted of three different tasks in which subjects had to make similarity judgments in response to visually presented pairs of concrete and imageable words. In one task, subjects had to indicate if the two words of each pair looked alike or shared several letters in common (visual similarity; e.g., plane–plant). In a second task, subjects had to indicate if the two words were semantically related. Semantic relatedness was based on cohyponymy: the related items were members of the same semantic category (semantic judgment; e.g., peach–plum). Finally, in a third task, subjects had to indicate if the two words sounded alike (rhyme judgment; e.g., boot–flute). This study is in many ways similar to previous studies concerning the functional lateralization of the brain for language in normal subjects and split-brain patients. According to the results of these studies, brain damage should have differential effects on performance in each of the three tasks depending on which hemisphere is affected. More specifically, a lesion of the right hemisphere should disturb the visual (i.e., orthographic) and semantic processing of words, whereas a lesion of the left hemisphere should interfere with the normal functioning of the phonological and semantic processes associated with language.

Chiarello and Church (1986) began the analysis of their results by comparing the performances of the left-brain–damaged, right-brain–damaged, and normal control subjects. The left-brain–damaged subjects made significantly more errors than the normal controls on the visual similarity task. However, there was no difference between the right-brain–damaged subjects and either the left-brain–damaged subjects or the normal controls. Results for the rhyming task revealed that the right-brain–damaged subjects made significantly fewer errors than the left-brain–damaged subjects and obtained scores similar to those of the normal controls. Finally, on the semantic judgment task, no significant difference was found between the two groups of brain-damaged subjects, though both groups produced more errors than the normal controls. Thus the right-brain–damaged subjects demonstrated an impairment in the ability to associate words on a semantic basis.

In the next phase of their analysis, Chiarello and Church (1986) compared the error percentages for each of the three groups of subjects on each of the three tasks. The results of this analysis revealed that the left-brain–damaged subjects had more difficulty in judging if the two words of a stimulus pair rhymed than if they were visually or semantically similar, these last two types of judgments being of a similar level of difficulty. The right-brain–damaged subjects obtained similar error percentages for the visually similar and rhyming words tasks, but they experienced considerably more difficulty on the semantic task.

Finally, Chiarello and Church (1986) took a closer look at the performance of their left-brain–damaged subjects. Although the small number of subjects did not allow for any statistical analyses, it was noted that the nonaphasic, left-brain–damaged subjects performed within normal limits on all three judgment tasks. According to Chiarello and Church, the relatively poor performance of the left-brain–damaged group can be attributed solely to the scores of the aphasic subjects. It seems then that as far as left-brain–damaged subjects are concerned, the presence of aphasia rather than of cerebral damage results in a disturbance of metalinguistic judgments. This analysis based on the heterogeneity of left-brain–damaged subjects suggests, still according to these investigators, that the difficulties of the right-brain–damaged subjects are not, a priori, due to a simple lesion effect. Chiarello and Church proposed instead that their difficulties are truly of a linguistic or semantic nature, at least as determined by their performance on the semantic judgment task. Pointing to the fact that none of their right-brain–damaged subjects suffered from any memory disorder, cognitive impairment, or mental deterioration, Chiarello and Church argue that the semantic impairment of right-brain–damaged subjects – or at least of those tested in the present study – is not attributable to a global cognitive disturbance. Furthermore, they add that metalinguistic difficulties in right-brain–damaged subjects are selective insofar as they are observed in the semantic task but not in the rhyming task. Chiarello and Church also point out that although they used a different procedure, the performances of their right-brain–damaged subjects were very much similar to those of Lesser's (1974) subjects, who demonstrated disorders of a semantic nature though they had no difficulty with the phonological processing of words. The results of these two studies converge toward the same interpretation: certain aspects of semantic processing are sustained by a bihemispheric neural organization. Therefore, it is highly likely that the right hemisphere contributes to the processes involved in metalinguistic judgments that are based on word meaning (Chiarello & Church, 1986).

In a comparison of their findings with those of Brownell et al. (1984, 1985, 1986), Chiarello and Church (1986) note that the semantic impairment of their own right-brain–damaged subjects was neither attributable to a deficit involving the connotative processing of words nor the result of a disproportionate amount of attention given to the literal meaning of the stimulus words. In fact, these authors emphasize that their semantic task consisted of concrete words and made use of the denotative meaning of these words. Chiarello and Church also draw attention to the fact that a number of studies have demonstrated the presence of verbal comprehension disorders in right-brain–damaged subjects (Gainotti et al., 1979, 1981, 1983; Lesser, 1974) using tasks that included semantic distractors. According to Chiarello and Church, it is likely that the difficulties of the right-brain–damaged subjects originate at the level of a postlexical judgment of the appropriate semantic nature of a response. Following a suggestion made by Gardner et al. (1983), Chiarello and Church suggest that the main linguistic impairment of right-brain–damaged subjects could involve the judgment of the plausible or appropriate nature of linguistic information. However, one should note that other studies, in which tasks requiring an evaluation of the plausible nature of the possible responses have been used, have shown normal performances in right-brain–damaged subjects (Bishop & Byng, 1984; Gainotti et al., 1979).

In contrast with the findings of Chiarello and Church (1986), Goulet, Joanette, Gagnon, and Sabourin (1989) reported that right-brain–damaged subjects are comparable to normals with respect to their sensitivity to cohyponymy. In their study, Goulet et al. investigated whether the nature of the lexicosemantic impairments observed in right-brain–damaged patients provides further support for Drews's (1987) hypothesis. As reported earlier in this chapter, Drews claimed that the two hemispheres support different lexicosemantic organization. The right hemisphere would predominantly allow for interconceptual relationships (contiguity), whereas the left would do so predominantly for intraconceptual relationships (similarity, e.g., cohyponymy). Consequently, Goulet et al. postulated that if Drews's hypothesis was right, the lexicosemantic impairments associated with a right-hemisphere lesion should affect conceptual relations based on the unity of perceivable scenes (interconceptual relationships) more than intraconceptual relationships.

Goulet et al. (1989) further sought to assess the specificity to the lexical level of these possible impairments. Thus, as already done with aphasic populations, Goulet et al. tested the sensitivity of right-brain–damaged subjects to relations between words as well as between pictures of objects. Pairs of related and unrelated objects (words or pictures) were used. Right-brain–damaged patients (N = 15) and normal controls (N = 15) were asked to indicate whether there was a relationship between items of each pair. Items paired on the basis of an intraconceptual relationship were more representations of the same semantic category (cohyponymy; e.g., vegetables) than of a common visual scene. The reverse was true for the items paired on the basis of an interconceptual relationship (e.g., cloud–airplane). Statistical analysis failed to reveal any significant difference between the performance of right-brain–damaged subjects and that of normal controls. Right-brain–damaged subjects were comparable to normals with respect to their sensitivity to intraconceptual and interconceptual relationships between words or pictures.

Many factors might explain the contrast that exists between the findings of Goulet et al. (1989) and those of Chiarello and Church (1986). One of them concerns the nature of the control group, which was constituted of hospitalized subjects in the study of Goulet et al. This factor seems to be responsible for generally lowered performances on cognitive tasks (Klonoff & Kennedy, 1966) and was not totally controlled by Chiarello and Church. A second possibility is that the right-brain–damaged subjects represent a nonhomogeneous population; some of these subjects may have few, if not any, lexicosemantic impairments, whereas others may present themselves with a lexicosemantic impairment, in which case it could be linked or not with the presence of a more general intellectual deficit (see Gainotti et al., 1979). In relation to this second factor, it should be noted that the exact role of the cognitive deterioration consecutive to some lesion of the right hemisphere should be investigated more closely. It seems clear that the actual existence of a lexicosemantic deficit in right-brain–damaged patients has to be confirmed and that the factors contributing to such deficit should be clarified. One should probably ask who the right-brain–damaged subjects suffering from a lexicosemantic deficit are rather than whether right-brain–damaged patients show some lexicosemantic impairment.

Role of the Effort Involved in the Processing of Semantic Judgments

In the second half of the eighteenth century, Baillarger and Hughlings-Jackson (see Ombredane, 1951) contrasted the most automatic use of words with the less automatic, or more voluntary, use of words in order to explain the linguistic behavior of aphasics. Today, aphasiologists hardly ever make reference to the writings of these pioneers. Be they right or wrong, they prefer to follow the lead of more modern researchers like Posner and Snyder (1975) and Neely (1977). Thus the distinction made between the automatic activation of information in memory and the conscious activation of the same information is presented as if it was a new finding pertaining to current cognitive theories. More and more authors, like Milberg and Blumstein (1981), reported over the last years that the semantic deficit found in certain types of aphasia is not absolute since many aphasics retain linguistic information about words that may be automatically activated although inaccessible to conscious processing during metalinguistic tasks.

Recently, following the movement initiated in aphasiology, some researchers have investigated the lexicosemantic deficit found in some right-brain–damaged patients in light of the dissociation between the automatic versus effortful (conscious or controlled) activation of semantic knowledge. For instance, the purpose of the study of Tompkins and Jackson (1988) was to examine the possibility that right-brain–damaged patients remain sensitive to lexical connotation at a relatively automatic level, although impaired at a more conscious, or effortful level, as reported by Brownell et al. (1984). Brain-damaged patients and control subjects were submitted to two auditory lexical-decision tasks. Each task involved a semantic priming condition: automatic versus effortful. With respect to the performances of right-brain–damaged subjects, Tompkins and Jackson (1988) reported facilitation effects without any interference effects (or inhibition) in the two priming conditions. These results suggest that right-brain–damaged patients remain sensitive to lexical connotation at a relatively automatic level. The authors concluded that the insensitivity to connotative meanings reported in right-brain–damaged patients by Brownell et al. (1984) should be attributed to deficiencies in the effortful processing of lexical connotation rather than to loss of knowledge of the connotative aspects of words.

The conclusions of Tompkins and Jackson (1988) are based on the absence of interference for the unrelated prime–target word pairs in the effortful condition. On a theoretical basis (see Neely, 1977), interference effect is expected when meaning activation is partially based on some effortful processing. However, in opposition to the interpretation of Tompkins and Jackson (1988), it must be noted that an absence of interference does not necessarily correspond to a lexicosemantic impairment. Other factors, such as the time needed to process (e.g., visually) the prime words or the understanding of the instructions, might also explain the absence of interference. Thus other studies are needed to validate the interpretation of Tompkins and Jackson.

The priming paradigm has also been used by Gagnon, Joanette, Goulet and Cardu (1988) to investigate the lexicosemantic deficiencies of the right-brain–damaged population. Semantic priming was induced by the use of pairs of words related on the basis of cohyponymy (e.g., desk–bed). Besides the automatic and controlled priming conditions obtained in the context of two lexical decision tasks, Gagnon et al. administered their subjects a semantic judgment task. This task involved cohyponyms as well as unrelated word pairs. Word stimuli of this task were translated from the protocol used by Chiarello and Church (1986). Right-brain–damaged subjects and normal controls were found to be comparable with respect to the priming effects (automatic and controlled).

However, facilitation and interference were found in both priming conditions. Right-brain–damaged subjects differed from normal controls only in the semantic judgment task.

Interference in the automatic priming condition was not expected since it could reflect some form of attentional (conscious or effortful) processing of the primes. Thus the performance on the automatic priming condition must be discarded. Nevertheless, performance in the controlled priming condition suggests that right-brain–damaged patients were able to undertake some form of conscious processing of word meanings. However, results on the semantic judgment task suggest that this capacity was limited and comply with a continuum of effort involved in the semantic processing of words required by the task.

In summary, the distinction made between the automatic and the effortful processing of word meanings highlights a new dimension that should be taken into account in trying to understand the effects of the nature of the lexicosemantic impairment in right-brain–damaged right-handers. Indeed, it seems that the lexicosemantic difficulties of right-brain–damaged subjects are the expression neither of an impairment of lexicosemantic knowledge nor of its automatic utilization or activation; rather they reflect an impairment of the effortful, and particularly of the most effortful, activation of the lexicosemantic information.

Conclusion

The results from studies involving semantic judgment tasks suggest that right-brain–damaged subjects have certain difficulties with respect to the processing of the antonymic, connotative, and metaphoric aspects of word meanings. Such subjects seem to have certain difficulties using the conceptual characteristics evoked by words in establishing semantic relationships. These observations suggest that in the right-handed individual, the presence of an intact right hemisphere is essential for the adequate semantic processing of words. However, the results of the various studies reviewed here are not unanimous. This lack of convergence could be due, at least in part, to methodological differences. For example, the subjects tested by Wilkins and Moscovitch (1978) had undergone a partial lobectomy of the anterior portion of the temporal lobe, and the presence of an epileptogenic focus since early childhood may have influenced the functional organization of the brain. Another example concerns the various levels of difficulty with respect to the type of semantic judgment required (cf. Goulet & Joanette, 1988). These methodological differences will require further investigation.

In addition, several of the studies reviewed here provided no information concerning the intellectual capacities of the right-brain–damaged subjects examined. It is, therefore, difficult to assess the influence of an eventual global cognitive deterioration, subsequent to cerebral damage, along with the lexicosemantic specificity of the right-brain–damaged subject's impairment, as well as the true nature of the right hemisphere's contribution to performance on tasks with lexicosemantic requirements. Finally, a new dimension should be taken into account in trying to understand the effects of the nature of the lexicosemantic impairment in right-brain–damaged right-handers: the distinction to be made between the automatic and the effortful processing of word meanings.

Word-Naming Tasks

Word-naming tasks correspond to the oral or written production of as many different words as possible, respecting a specific criterion, within a time period that usually varies for 60 to 120 seconds. Two types of criteria can be used. On the one hand, formal or orthographic criteria involve a feature of the surface structure of the words to be evoked (e.g., words beginning with the letter *b*). On the other hand, semantic criteria involve one or several semantic characteristics of the words to be produced (e.g., animal names).

Milner (1964) instructed epileptic patients who had undergone either a right or left prefrontal lobectomy, or an excision of the left temporal lobe, to write down as many words as possible beginning with the letter *s*. The subjects were allowed five minutes to perform the task. Then the subjects were given four minutes to write down as many four-letter words as possible beginning with the letter *c* (Thurstone's task). Results showed that only the left-brain–damaged patients with frontal lesions were impaired on these tasks, whereas patients with right frontal lesions performed normally.

Benton (1968) administered the FAS test to nonaphasic right-handed subjects with either right or left frontal lesions or bilateral lesions. In the FAS test 60 seconds is allowed for subjects to orally produce words beginning with each of the three letters *f*, *a*, and *s*. Results showed that right-brain–damaged subjects provided more words than either of the other two groups of subjects. However, 38% of the right-brain–damaged subjects and 70% of the left-brain–damaged subjects as well as of the subjects with bilateral lesions scored lower than 95% of the normal controls. Benton, like Milner (1964), concluded that the effects of frontal damage on lexical availability were more important in subjects with left or bilateral lesions than in subjects with right-sided lesions, even when the presence or absence of aphasia was taken into account. In light of his own findings, as well as those of Milner, Benton suggested that the deficit caused by left-brain damage affects higher levels of linguistic functioning since the manifestations are independent of the modality used (i.e., oral or written). Finally, Benton did not retain the existence of a disorder in right-brain–damaged subjects, in spite of the relatively poor performance of some of these subjects.

In a similar study, Ramier and Hécaen (1970) instructed brain-damaged subjects to orally produce as many words as possible beginning with the letters *p*, *f*, and *l* (60 seconds per trial). Six groups of brain-damaged subjects differing in terms of lesion site (i.e., right or left, frontal or not) and the presence or absence of aphasia were included in the study. Only factors related to lesion site were found to have a significant bearing on the results. Ramier and Hécaen observed that subjects with left frontal lesions produced the fewest number of words, thereby confirming the findings of Milner (1964) and Benton (1968). However, Ramier and Hécaen reported that verbal productivity was also affected by a right frontal lesion, although not as severely as by a left frontal lesion. According to the researchers, the integrity of the frontal lobes is essential for normal "spontaneity" and/or "initiative." Damage to either of the frontal lobes results in a lack of initiative and aspecific aspontaneity, which is more pronounced for the functional domain normally sustained by the damaged hemisphere. It is therefore normal that a frontal lesion of the right hemisphere results in the reduction of lexical availability. While such a reduction is less important than that observed following a left frontal lesion, it is nonetheless present. This does not necessarily mean, however, that the minor hemisphere has specific lexicosemantic abilities.

The involvement of a frontal component, rather than a linguistic component, to explain the reduced lexical availability in frontal brain-damaged subjects has also been examined by Perret (1974). Right-handed subjects with unilateral right- or left-hemisphere lesions as well as a group of normal control subjects were given an oral version of Milner's (1964) task. The groups of subjects were similar with respect to mean age and IQ (>70), while the brain-damaged groups also shared similar lesion etiologies (i.e., neoplasms for the most part). Perret found a greater reduction of verbal fluency in subjects with frontal lesions compared to subjects with temporal or posterior lesions, regardless of which hemisphere was affected. The results also showed that right-brain–damaged subjects, as a group, were less fluent than normals. This verbal reduction was particularly evident in subjects with frontal lesions. Perret attributed the impaired performance of the frontal subjects to a difficulty in adapting themselves to an unfamiliar situation, rather than to a disorder of spontaneity and/or initiative (see Ramier & Hécaen, 1970). According to Perret, the use of letters as a production criterion creates an unusual situation since the search for words is normally conducted on a semantic basis. It is therefore likely that tasks with such criteria evaluate the capacity to suppress the usual behavior of looking for words as a function of their meaning rather than simply looking for words. Perret found further evidence to support this hypothesis in the fact that the productions of subjects with right or left frontal lesions were correlated with their performance on a modified version of the Stroop test.

More recently, Bruyer and Tuyumbu (1980) used a task similar to the one used by Ramier and Hécaen (1970). They reported that the decrease in lexical availability following unilateral cerebral damage did not differ significantly from one hemisphere to the other, even when the performances of subjects with mild aphasia were taken into account. A frontal lesion, however, resulted in a disorder that was significantly more important than a nonfrontal lesion. These investigators also noted that certain nonfrontal right-brain–damaged subjects were judged to be deficient (i.e., when compared with normals relative to a preestablished cutoff score), whereas none of the nonaphasic, nonfrontal left-brain–damaged subjects were classified as such. The results of this study suggest that a lesion of the right hemisphere can impair task performance, although this reduction cannot be explained on the sole basis of a lesion effect since none of the scores of the nonaphasic left-brain–damaged subjects was abnormally reduced. It could be, however, that differences in lesion size were responsible for this apparent difference between nonaphasic right- and left-brain–damaged subjects. Without claiming that this factor does indeed account for these results, it is important to keep in mind that it is very difficult to come up with a group of nonaphasic left-brain–damaged subjects in whom lesion size is comparable to that observed in an unselected group of right-brain–damaged subjects. Although Bruyer and Tuyumbu did not provide any indications concerning lesion size, the role that this factor may have played in this study cannot be overlooked.

The verbal fluency task devised by Ramier and Hécaen (1970) was also used in an experiment conducted by Cavalli et al. (1981). However, they did not take lesion site into consideration. Nonetheless, word-naming performance was adjusted for the age and schooling of the subjects. Results showed no significant difference between right-brain–damaged subjects and normals. In light of these results and those obtained in other linguistic tasks, Cavalli et al. concluded that their right-brain–damaged subjects demonstrated no oral language impairment of the types encountered subsequent to lesions of the classic language area. In a similar vein, Bentin and Gordon (1979) reported that right-brain–damaged subjects, with frontal or central lesions for the most part, obtained near-normal scores when they had to write down as many words as possible beginning with a particular letter. However, there was tendency for right-handed, right-brain–damaged subjects to produce fewer words than normal controls (i.e., 27.78 words and 33.27 words respectively).

Both lesion site and size were taken into consideration in a study conducted by Miceli, Caltagirone, Gainotti, Masullo, and Silveri (1981). The subjects were administered a task similar to the one used by Benton (1968). None of the subjects were aphasic. However, the authors neglected to indicate whether all of their subjects were right-handed. Be that as it may, results for the subjects with a lesion confined to a single lobe revealed that left-brain–damaged subjects were significantly less productive than right-brain–damaged subjects, regardless of the lobe affected. As for subjects with a lesion involving more than one lobe, the performance of right-brain–damaged subjects was most affected when the frontal lobe was damaged. Bornstein (1986) reported similar findings. More specifically, left-brain–damaged subjects produced fewer words than right-brain–damaged subjects, while subjects with frontal lesions produced fewer words than those with nonfrontal lesions. According to Bornstein, left-brain–damaged subjects with nonfrontal lesions and right-brain–damaged subjects with frontal lesions performed better than subjects with left frontal lesions, who produced the fewest number of words, but they did not perform as well as subjects with nonfrontal right lesions, who produced the greatest number of words. Unfortunately, neither of the studies included a group of normal control subjects, which makes it virtually impossible to determine whether the performance of the nonfrontal right-brain–damaged subjects was normal.

Pendleton, Heaton, Lehman, and Hulihan (1982) administered the Thurstone task to brain-damaged subjects with various lesional sites and found that the reduction of verbal fluency was indeed more important in subjects with frontal lesions and in left-brain–damaged subjects compared to right-brain–damaged subjects. This latter difference was due to the greater reduction of verbal fluency in subjects with left frontal lesions compared to those with right frontal lesions. According to the authors, verbal fluency is sensitive to any form of cerebral damage. Along the same lines of thought, Albert and Sandson (1986) observed that normal subjects produced more words on an oral version of the FAS than did right-brain–damaged subjects in whom the exact lesional site was unknown.

An important factor to consider when using a word-naming task is the productivity coefficient of the criterion involved. Thus depending on whether there is a large number or a small number of words beginning with the same letter, it could be either more easy or more difficult to observe a verbal reduction subsequent to cerebral damage (Bolter, Long, & Wagner, 1983; Borkowski, Benton, & Spreen, 1967). Intrigued by this factor, Bolter et al. sought to determine if the productivity coefficient for the letters *s* and *c* in the Thurstone task had an effect on the task's clinical value. The methodological constraints of Thurstone's task imply that *c* is less productive than *s*, since the subjects are given five minutes to produce four-letter words beginning with the letter *s* but only four minutes to produce words beginning with the letter *c*. Bolter et al. also examined the relationship between the verbal intelligence of brain-damaged subjects and the clinical value of Thurstone's task. Assuming that a lesion of the right hemisphere is associated with a verbal reduction, the results showed that this reduction is more apparent in subjects whose verbal IQ is less than 100 and for criteria with a high productivity coefficient (e.g., the letter *s*).

The use of formal or orthographic criteria has failed to provide any consistent evidence to the effect that the right hemisphere sustains a lexicosemantic contribution to word-naming performance. The results obtained with semantic criteria are just as ambivalent. For example, Newcombe (1969) examined the oral production of animal names and objects (60 seconds per criterion). She found that the scores of right-brain–damaged subjects were similar to those of normal subjects and that this absence of any significant difference seemed to be maintained independently of the lesion site (frontal, parietal, etc.) or the criterion. In contrast, Boller (1968) reported that after having corrected subjects' scores for age and schooling, brain-damaged subjects (right and left) produced significantly fewer animal names than normal controls. Moreover, Boller was unable to find any significant difference between right-brain–damaged subjects and nonaphasic left-brain–damaged subjects, in spite of the fact that scores were corrected for lesion size as estimated by reaction times. According to Boller, however, the impairment of both right- and left-brain–damaged subjects reflects a nonspecific consequence of cerebral damage rather than a disturbance of a linguistic nature. Boller goes on to suggest that word-naming tasks involve linguistic capacities at a level where they are strongly directly related to higher order processes such as intelligence or a particular form of mental energy.

In contrast to the previous studies, the results of an experiment conducted by Lifrak and Novelly (1984), with subjects who had undergone a right or left temporal lobectomy, suggest that the word-naming impairment of right-brain–

damaged subjects is linked to the presence of semantic constraints rather than to an impairment involving word recovery and/or production. The subjects were administered a task that involved either the simple production of words with no sentence formation (i.e., "Word Fluency") or the production of words in response to semantic criteria (i.e., "Set Test"). In both tasks, the scores of right-brain-damaged subjects were similar to those of left-brain-damaged subjects. However, when their performance was compared with that of normal subjects, the right-brain-damaged subjects produced just as many words on the Word Association task but significantly fewer words on the Set Test.

Grossman (1981) also used semantic criteria (e.g., sports, birds, arms; 60 seconds per criterion) to study word-naming performance in brain-damaged subjects. However, Grossman was not only interested in the quantitative aspects of word-naming performance but was also one of the first to conduct a qualitative analysis of the performance of left- and right-brain-damaged subjects on a word-naming task that involved semantic criteria. At the quantitative level (i.e., mean number of words produced for each criterion), Grossman observed a significant difference between aphasics (i.e., fluent or nonfluent) and right-brain-damaged subjects or normal controls. No differences were observed between these last two groups. At the qualitative level, Grossman reported that the performance of right-brain-damaged subjects was similar to that of normals in terms of the typicality index as well as in terms of the various degrees of typicality covered by the words produced. The typicality index refers to the degree to which a given element is representative or prototypical of the semantic category to which it belongs. For example, the word "robin" is more prototypical of the category "birds" than is the word "penguin."

Grossman (1981) also reported that in right-brain-damaged subjects, a greater proportion of the items produced were grouped into clusters, although there was no difference between the three groups of subjects with respect to the mean number of elements contained within each of these clusters. For the purposes of the present study, a cluster was said to be created when at least three consecutively produced items shared at least one or several "semantic" features. A particularly interesting finding was that the three groups of subjects did not produce the same number of clusters. The left-brain-damaged subjects produced few clusters, but the items included had many overlapping features, the right-brain-damaged subjects produced many clusters with the items showing little overlap, and the performance of normal subjects fell between these two extremes. In light of these findings, Grossman concluded that right-brain-damaged subjects were very analytic, searching for attributes shared by several items. On the other hand, left-brain-damaged subjects were less analytic since their clusters were based on common functional and perceptual characteristics that were much more obvious. Grossman leaves the impression that he interprets the strategies adopted by both groups of brain-damaged subjects as being compatible with the classic analytic-holistic dissociation that was postulated to explain the functional lateralization of the brain.

To demonstrate that the right hemisphere's contribution to the processing of words is purely semantic, Joanette and Goulet (1986a, 1988) used both formal and semantic criteria. Their results show that the mean number of words produced by right-brain–damaged subjects, in response to formal criteria (i.e., words beginning with the letters *b* and *r*; production time: 2 minutes per criterion), was not significantly different from the performance of normal hospitalized subjects, regardless of whether one considered the total number of words produced or if the unacceptable words (i.e., nonwords, repetitions, nonrespect of criteria) were excluded. As for the semantic criteria (i.e., animal names and furniture names; production time: 2 minutes per criterion), Joanette and Goulet did observe a quantitative difference between the performances of right-brain–damaged and normal subjects, regardless of whether one considered all of the words produced or only the acceptable words. Furthermore, the number of words produced by frontal and nonfrontal right-brain–damaged subjects was similar, regardless of the specific criteria used. These observations suggest, according to Joanette and Goulet, that the verbal reduction of right-brain–damaged subjects is not the mere reflection of a frontal impairment addition; the eventual contribution of the right hemisphere to linguistic functioning appears to be more important for the semantic aspects than for the formal aspects of language. Such an interpretation is compatible with the results from studies of split-brain subjects.

In an attempt to further describe the nature of the impairment of their right-brain–damaged subjects, Joanette, Goulet and Ledorze (1988; see also Joanette and Goulet, 1988) conducted a qualitative analysis of their subjects' word-naming performance based on semantic criteria. In so doing, they found that it was only after the first 30 seconds of production that the right-brain–damaged subjects began to produce fewer words than the normals. Furthermore, they found no significant difference between right-brain–damaged subjects and normals with respect to the number produced, the nature, or the distribution of unacceptable words. These observations, combined with those of the preceding study, led Joanette et al. to conclude that the difficulty of right-brain–damaged subjects is (1) not aspecific or perseverative in nature; (2) more of a lexicosemantic and particularly semantic nature; and (3) linked to the most active or effortful phase of production, which requires a more controlled memory search compared with the first 30 seconds of production, during which the activation of lexical and/or semantic organization is more automatic.

In the reanalysis of their results, Goulet and Joanette (1987; see also Joanette and Goulet, 1988) examined in detail two strategies believed to facilitate lexical evocation from semantic criteria and sought to verify if right-brain–damaged subjects differed from normals with respect to the use of these strategies. The rationale is that optimal word-naming performance for semantic criteria (e.g., animals) is achieved by (1) the evocation of as many semantic subfields as possible (e.g., farm animals, African animals, fish) and (2) the maximum exploitation of these various subfields. Goulet and Joanette noted that when the animal names criterion was used, right-brain–damaged subjects used fewer semantic subfields than normals[8] and produced fewer words[9] within the most exploited subfield. In addition, Goulet and Joanette reported that right-brain–damaged subjects, as a group, produced fewer different words (143) than normal subjects (174), in spite of the fact that a greater number of right-brain–damaged subjects participated in the study (i.e., 35 as opposed to 20 normals). These results suggest that the two strategies postulated are impaired following a lesion of the right hemisphere. They are also compatible with the hypothesis of a lexicosemantic disorder in right-brain–damaged subjects and are favorable to a contribution of the right hemisphere to the semantic aspects of lexicosemantic functioning.

In the wake of these findings, Goulet and Joanette (1987) remain somewhat cautious. They note, for example, that the specificity (i.e., specific to the right hemisphere and/or semantically specific) of an eventual semantic contribution of the right hemisphere is still an unanswered question. Caution is also warranted in light of several methodological limitations: the absence of nonaphasic left-brain–damaged subjects or subjects with subcortical lesions that would allow an evaluation of the influence of a lesion effect; the absence of any information concerning the general intellectual capacities of the subjects, in spite of the fact that the two groups of subjects were similar in terms of age, sex, schooling, and manual preference (all the subjects were right-handed); absence of any information concerning the productivity coefficient of the criteria used (Goulet & Joanette, 1987).

Laine (1987) and Laine and Neimi (1988) compared the word-naming performance of right-brain–damaged and nonaphasic left-brain–damaged subjects. Their findings support the conclusions of Joanette and Goulet (1986a, 1987; Goulet & Joanette, 1987). More specifically, the investigators report that right- and left-brain–damaged subjects with vascular lesions, as well as neurologically intact subjects, produced a similar number of words beginning with the letter *s* (production time: 60 seconds). However, right-brain–damaged patients produced significantly fewer animal names than normals, although their performance was comparable to that of nonaphasic left-brain–damaged subjects. A more detailed analysis of the subjects' performance revealed that the size of the clusters of animal names (i.e., names sharing a number of semantic features) was smaller in left-brain–damaged subjects than in right-brain–damaged subjects or normals. Conversely, right-brain–damaged subjects produced fewer semantic clusters of animal names and used fewer animal subcategories (e.g., birds, fish) than left-brain–damaged subjects or normals. These observations suggest that the impairment of right-brain–damaged subjects is qualitatively different from that

[8]A semantic subfield is said to occur when at least three consecutively produced words share one or several semantic features in common. These features form the label of the subfield and are used to group as many consecutively produced words as possible.

[9]In tabulating the total number of words belonging to a particular semantic subfield, the fact that the related words were consecutive or not was disregarded, insofar as the subfield was judged to be formed on the basis of three consecutively produced words.

observed in nonaphasic left-brain–damaged subjects, a finding that goes against the hypothesis of an aspecific lesion effect. The authors conclude that in production situations, right-brain–damaged subjects have certain difficulties with the use of semantic strategies.

Diggs and Basili (1987) instructed right- and left-brain–damaged patients and normal controls to produce as many animal names as possible (production time: 2 minutes) and then to name anything that could roll. The subjects were also administered a picture-naming task and then asked to describe the function of each of the illustrated objects. Diggs and Basili report that the control subjects obtained higher scores than either of the groups of brain-damaged subjects. The right-brain–damaged subjects performed better than the nonaphasic left-brain–damaged on the two word fluency tasks. The two groups of brain-damaged subjects differed with respect to the total number of words produced but also with respect to the number of different categories (e.g., animals' names: frog–amphibians, eagle–birds, etc.) present in the response set. There were no differences in the performances of the two groups of brain-damaged subjects on either the picture-naming or object-function task. Interestingly, however, Diggs and Basili report that nearly all of the errors of the right-brain–damaged patients on the picture-naming task were either semantic or the result of failure to respond. According to these investigators, right-brain–damaged patients do have communication problems or verbal production difficulties, whose underlying cause or causes could be linguistic, cognitive, or both.

Unlike these last studies, Cappa, Papagno, and Valler (1987) reported that all of their right-brain–damaged subjects obtained scores within the range of those of normal controls, regardless of the type of word-naming criterion used (i.e., semantic or formal). They conclude that in right-handed subjects, the presence of an intact left hemisphere is all that is needed to ensure the execution of several linguistic tasks.

Sabourin, Goulet, and Joanette (1988) recently provided evidence to corroborate Cappa's observations. An important methodological flaw in the Joanette and Goulet (1986a) and Laine and Niemi (1988) studies was the failure to take into consideration the productivity coefficient for the formal and semantic criteria that were used. Sabourin et al. employed a word-naming task in which orthographic and semantic criteria were chosen from three different levels of productivity (i.e., high, medium, low). The results of this study show that right-brain–damaged subjects produced significantly fewer words only when the two types of criteria had a high level of productivity. Contrary to the hypothesis formulated by Joanette and Goulet and Laine and Niemi, it would appear that the semantic or formal nature of the criteria is not a determining factor in the impaired word-naming performance of right-brain–damaged patients. In previous studies, where it was shown that right-brain–damaged patients obtained lower scores for semantic criteria but not for formal criteria, it may have been that the semantic criteria had a higher productivity coefficient than the orthographic criteria. Thus Sabourin et al.'s results do not seem to demonstrate any particular role of the right hemisphere with respect to the purely semantic aspects of lexical semantics.

In summary, the use of formal or semantic criteria in word-naming tasks does not always provide evidence in support of the hypothesis that the integrity of the right hemisphere is necessary for the recovery and/or the production of words. When the evidence does happen to be favorable to such a hypothesis, it suggests above all a contribution of the frontal and prefrontal regions of the right hemisphere to nonspecifically linguistic or semantic aspects of behavior. The fact that nonfrontal damage of the right hemisphere results in a decrease of lexical availability, without the underlying deficit being simply an aspecific lesion effect, is not being unanimously accepted. Many studies should be reconsidered in light of the involvement of a number of factors, such as overall intellectual functioning or the level of difficulty of the criteria, in order to better understand the discrepancies among the various findings published to date. Qualitative analyses of word-naming performance involving semantic criteria are more favorable to the hypothesis of a lexicosemantic contribution of the right hemisphere. Further studies are required to provide a more detailed evaluation of this hypothesis. However, one thing should be kept in mind when using the qualitative approach: to estimate the real contribution of the right hemisphere to lexical semantics, one must determine not only if right-brain–damaged subjects succeed at a given task but also how right-brain–damaged patients perform such a task and what strategies they use.

Studies of Right-Brain–Damage: Conclusion

Several indications suggest that a lesion of the right hemisphere in a right-handed individual can impair performance on tasks of a lexicosemantic nature. Apart from the difficulties that are secondary to a lesion of the frontal lobes or to the presence of visuoperceptual disorders, there are three factors that could explain such an impairment. The first concerns the presence of a reduction in global intellectual efficiency. In this case, the impairment is not specifically of a lexicosemantic nature. Strict control of this factor does raise a problem, though. On the one hand, how does one assess the intellectual deterioration of a right-brain–damaged subject with nonverbal tools when contemporary neuropsychology preaches that the right hemisphere plays an important role with respect to nonverbal processes? On the other hand, how does one assess the intellectual deterioration of a right-brain–damaged subject with verbal tools if, according to the hypothesis under examination, the right hemisphere contributes to linguistic processes? Furthermore, to what extent is it utopian to try to dissociate semantic functioning from global intellectual functioning?

Secondly, it may be that the lexicosemantic impairment of right-brain–damaged subjects is attributable to an aspecific lesion effect. This hypothesis does not entirely eliminate the possibility of a specific lexicosemantic contribution under the control of the right hemisphere. Rather, it emphasizes the vulnerability and the complexity of lexicosemantic processes. Functional double dissociations (e.g., Brownell et al., 1984) are conducive to the hypothesis that the right hemisphere has a lexicosemantic contribution of its own. However, these double dissociations have yet to be replicated or found for other aspects of lexicosemantic functioning.

Finally, it may well be that the right hemisphere does indeed make a specific contribution to lexicosemantic processes. The preceding pages have shown that this hypothesis is quite plausible, though it requires further study.

Conclusion

Studies of commissurotomized and normal subjects have shown that the isolated right hemisphere is capable, under certain instances, of subtending a certain lexicosemantic potential. Beyond this lexicosemantic potential, results from numerous studies of right-brain–damaged subjects support an actual contribution of the right hemisphere to lexical semantics. However, several studies have produced evidence to the contrary, thereby underlining the difficulties, both theoretical and methodological, that one must overcome in order to ensure (1) the existence of a semantic potential and/or contribution that is specifically under the control of the right hemisphere in the right-handed individual and (2) the extent to which this potential and/or contribution is specific to language. Despite the large number of studies published to date, more methodologically and conceptually sound studies are needed to evaluate the effective actual role of the right hemisphere in the semantic processing of words.

5
Right Hemisphere and Written Language: The "Deep Dyslexia" Case

This chapter is not aimed at providing a detailed review of the various visuoperceptual capacities of the right hemisphere that are involved in the processing of a particular type of visual material, namely, letters. However, it should be kept in mind that letter identification is possible in both hemispheres, though the involvement of the right hemisphere becomes more important as perceptual complexity increases (Beaumont, 1982a; Bradshaw & Nettleton, 1983). For example, in a study of the capacity to name letters presented in ten different fonts, Bryden and Allard (1976) found a right visual field superiority (left hemisphere) for printed letters but a left visual field superiority (right hemisphere) for handwritten letters. A superiority of the right hemisphere has also been found when the letter classification task was more complex (Jonides, 1979), when the letters were smaller in size (Miller & Butler, 1980), and when the letters were presented briefly (Bradshaw, Hicks, & Rose, 1979; Hellige & Webster, 1979). None of these studies has provided any information concerning the linguistic capacities of the right hemisphere for written language per se. They have shown only that the identification capacities of the right hemisphere are sometimes superior to those of the left hemisphere, particularly when the perceptual conditions are complex. As we will see, this capacity has been explored in studies in which very brief stimulus presentations were used (e.g., Landis, Regard, & Serrat, 1980). We will not examine the reading or writing difficulties, be they visuospatial or attentional in nature, of right-brain–damaged patients. The tendency to use only the right half of a sheet of paper, to space letters, to disregard the horizontal axis when writing, and to make downstroke duplications have been described by Hécaen and Marcie (1974). Reading errors due to left visual hemineglect (e.g., omitting the beginning of a word) have been described by Kinsbourne and Warrington (1962) as well as Baxter and Warrington (1983).

This chapter is organized around a central hypothesis, whose origins lie in a rather optimistic interpretation of some of the facts reported in the previous chapter concerning the lexicosemantic capacities of the right-hander's right hemisphere. The hypothesis in question upholds the right hemisphere's responsibility in the semiology of deep dyslexia, a particular type of reading disorder. This issue could have been examined in the chapter on the right hemisphere's contribution to aphasia. In reality, it would seem that this hypothesis—which has alternatives—can be used to illustrate several aspects of the relationship between the right hemisphere and written words. Coltheart (1980, 1983) formulated this hypothesis on the basis of the convergence between the semiology of deep dyslexia and the evidence from studies of normal subjects and commissurotomized patients, which concern the right hemisphere's potential to process written words. It is not our intention to review all of the problems associated with reading disorders which have been the object of recent publications (e.g., Ellis, 1984; Henderson, 1984). However, some information will be provided to illustrate the various levels of discussion that have arisen as a result of (1) the convergences as well as the contradictions of findings from studies of different populations, and (2) the confrontation between arguments that are different in nature, which are based either on the mechanisms supposedly involved in reading behavior or on more general concepts such as interhemispheric inhibition or cooperation.

Deep Dyslexia and the Right-Hemisphere Hypothesis

Following a lesion of the left hemisphere (which in most instances is massive), some patients [Broca's aphasics for the most part, but not exclusively (e.g., Kremin, 1980)] demonstrate a reading disorder that has been referred to as deep dyslexia, which includes the following features (Coltheart, 1980, 1983; Kremin, 1984):

1. The reading errors consist of semantic paralexias [e.g., "cercle" ("circle") is read as "rond" ("round")],[1] visual paralexias [e.g., "coude" ("elbow") is read as "coudre" ("to sew")], and derivational paralexias [e.g., "journée" ("day") is read as "journal" ("newspaper")].
2. Reading nonwords is impossible.
3. Functional words result in many more errors than content words; these errors consist in the substitution of a function word for another function word [e.g., "toi" ("you") is read as "nous" ("us")].
4. Reading concrete words with high imagery values is easier than reading abstract words with low imagery values.

The terms of the right-hemisphere hypothesis (Coltheart, 1980, 1983; Coltheart, Patterson, & Marshall, 1987) can be summarized as follows. First, it must be assumed that in cases of deep dyslexia, an extensive left-hemisphere lesion strongly interferes with orthographic processing, to the extent that written language is processed only in the right hemisphere. Therefore, the hypothesis for deep dyslexia is that lexical access and comprehension of the written code are carried out in the intact right hemisphere. It is also assumed that under these conditions oral responses remain under the control of the left hemisphere,

[1]The examples are taken from a study of patient S.P. by De Partz (1986).

which uses information that has been processed and transferred by the right hemisphere.

In the following paragraphs we examine the various implications of the terms of this hypothesis, based on its discussion by Patterson and Besner (1984a, 1984b). In addition, we go beyond the limits of deep dyslexia to examine the eventual role of the right hemisphere in the reading performance of aphasics. Although this chapter is critical of the studies suggesting a major responsibility of the right hemisphere in cases of deep dyslexia or in the production of semantic paralexias, it acknowledges the fact that there is indeed some evidence in support of a partial contribution of the right hemisphere.

Lateral Differences in Deep Dyslexics.

Ideally, evidence of the right hemisphere's contribution to the semiology of deep dyslexia could come from the modification of reading abilities or disabilities following a second lesion of the right hemisphere. Until now, the only available evidence has been indirect, coming essentially from tachistoscopic studies of a limited number of patients. Investigations of this type with brain-damaged patients (Beaumont, 1983) require that a number of methodological conditions be fulfilled to ensure that the lateral differences observed are not actually due to any perceptual disorder. The three patients examined by Saffran et al. (1980) and the methodology used in their study meet the necessary requirements. On the one hand, these patients had no visual field disorders. On the other hand, the duration of the stimulus presentations was adjusted for each hemifield using a letter-identification control task.

The experiment consisted of a lexical decision task during which 40 concrete words and 40 nonwords were presented in each hemifield. The results for two of the patients (V.S. and B.L.) revealed a slight superiority of the left visual field (right hemisphere). As for the third patient (H.T.), performance in each hemifield was inferior to that of the other two patients, with a very slight superiority of the right visual field (left hemisphere). Since the data did not result in acceptance or rejection of the hypothesis, Saffran et al. (1980) designed a complementary task in which the two hemispheres were forced to compete with each other. This competition was achieved using bilateral tachistoscopic presentations in which one word or nonword was presented in one hemifield while a letter was presented in the opposite hemifield. Under these conditions, the number of correct responses was higher for left visual field stimuli (right hemisphere) in all three patients. Although this advantage supported the hypothesis, it varied from one patient to another. Therefore, a contribution of the intact right hemisphere was possible, though the actual importance of this possible contribution was still not established. Saffran et al. were therefore cautious in their interpretation: ". . . the results would seem to indicate that the processing of graphemic information in deep dyslexia involves, to a varying, but significant degree, the intact right hemisphere" (1980, p. 395).

More recently, Zaidel and Schweiger reported the case of a 37-year-old patient (R.W.), presenting deep dyslexia with dysgraphia, who was examined eight months after suffering a left sylvian occlusion (Schweiger, Fiel, Dobkin, & Zaidel, 1983; Zaidel & Schweiger, 1983, 1984). The patient's visual field was normal. In a lexical decision task, the percentage of correct responses was identical for both hemifields (80%), but reaction times were significantly shorter for words presented in the left visual hemifield (right hemisphere). This difference did not seem to be the result of a perceptual disorder, since in a control task which consisted of determining the number of letters in a word (more than four letters or not) no difference was found in reaction times and the success rate was 99%.[2] The significant difference in reaction time suggests that the patient's right hemisphere was more involved in lexical decision than the left.

To summarize, the evidence from hemifield studies of four deep dyslexic patients suggests that there may indeed be a contribution of the right hemisphere to the reading of isolated words.

Phonological Recoding

The term "phonological recoding" seems to be preferred to "grapheme–phoneme conversion," which is used by some, because the mechanism of the shift from print to phonology, for nonwords or for irregular words, does not seem to be based solely on a grapheme to phoneme conversion (Young, Ellis, & Bion, 1984). Other mechanisms possibly involved include the similarity or the analogy with other words of the language (Glushko, 1979). Our intention here is to present evidence of the right hemisphere's inability to shift from print to phonology, without discussing the different mechanisms that might govern this recoding.

The inability of deep dyslexics to read nonwords aloud or to associate rhyming nonwords (Coltheart, 1980; Patterson, 1981) attests to their inability to perform phonological recoding. The hypothesis that reading performance in deep dyslexic patients depends on the right hemisphere is based on the notion that in commissurotomized patients, the right hemisphere is actually unable to perform this conversion from print to phonology. Evidence from several studies of normal subjects has led to a similar conclusion, although there is no unanimity.

Phonological Recoding and the Right Hemisphere of Commissurotomized Patients

As mentioned in Chapter 4, the right hemisphere of certain commissurotomized patients (Gazzaniga 1983a, 1983b; Levy 1983a; Myers 1984; Zaidel 1983b) is capable of connecting a word presented in the left hemifield with the correspond-

[2]The high rate of success may have been due to the relative easiness of the task (i.e., a ceiling effect).

ing picture from among several choices (Gazzaniga & Sperry, 1967). How this word–picture connection is carried out must be determined. The findings from the next group of studies agree that reading a word is probably lexical, since the right hemisphere is incapable of phonological recoding. Although Levy and Trevarthen's study (1977) provides a partial solution, it is essentially the work of Zaidel and Peters (1981) that has clearly demonstrated this incapacity of phonological recoding.

Levy and Trevarthen (1977) examined four commissurotomized patients (L.B., N.G., A.A., and C.C.). When a word was presented in the left visual field, these patients were able to select the corresponding picture. When presented with a chimerical word (e.g., the sequence "deon," in which "de" is presented in the left hemifield and "on" in the right hemifield) which had to be matched with one of three words (e.g., "deed," "noon," or "sees"), all of the patients, with the exception of C.C.,[3] chose the word corresponding to the left visual field (i.e., "deed"). This study revealed that in a task where only word matching was required, the competition between the two hemispheres favored the left hemifield, that is, the right hemisphere.

In another task the subjects were required to match chimerical words with one of three pictures. In contrast to the preceding task, the subject had to extract the meaning of the written word, since ideographic decoding was insufficient. Here patients favored the hemiword presented in the right hemifield. Stated otherwise, the right hemisphere is dominant only when the task calls for word matching (i.e., global processing), but as soon as linguistic processing is required, the left hemisphere takes over. These experiments show that the right hemisphere is capable of matching a word projected in the left hemifield with a picture, but that reading is performed in a global, lexical manner.

To demonstrate that the right hemisphere is incapable of phonological decoding, Levy and Trevarthen examined the capacity of commissurotomized patients to match rhyming words.[4] However, because they were using chimerical pictures and were not directly testing the actual conversion from print to phonology, this experiment could provide evidence that was partially compatible with the hypothesis only. The subjects were shown a chimerical word consisting of two half-words such as half of the word "rose" in the left field (ro) and half of the word "eye" (ye) in the right field. Their task was to choose from among three alternatives the picture associated with the word that rhymed with the target word (e.g., "toes" rhymed with "rose," "pie" with "eye," whereas "key" rhymed with neither). The subjects (with the exception of C.C.) showed a preference for the stimulus presented in the right hemifield. Thus they solved this rhyming task with their left hemisphere (i.e., they chose the picture of a pie). A more simple way of testing this incapacity of the right hemisphere would be to present entire pictures in the left hemifield. However, the patients still failed the task. Finally, their failure is similar when the matching task involves three orally presented words.

---

[3]C.C. was suspected of having an early massive lesion of the left hemisphere.

[4]This experiment included the four patients mentioned earlier as well as patient R.Y.

Zaidel and Peters (1981) sought to provide a detailed description of this incapacity in two commissurotomized patients (L.B. and N.G.) using several tasks that inevitably required phonological decoding: matching on the basis of homonymy, rhyme, or reading nonwords with the isolated right hemisphere. In the first task, the subject had to indicate two homonymous pictures among an array of four (e.g., an animal "bat" and a baseball "bat"). The other two pictures included a semantic distractor (e.g., "spider") and a phonological distractor (e.g., "hat"). A control task was used to ensure that each picture shown separately could be matched with its oral form. Both patients were able to perform the task successfully when the pictures were presented in the right visual hemifield (left hemisphere). However, their performance differed when the pictures were presented in the left visual hemifield. Whereas N.G. failed and favored semantic associations, L.B. was able to match the homonymous pictures, performing better than at the level of chance. In the second task, the two matching pictures had rhyming names (e.g., "bat" and "hat"), or phonetically similar names (e.g., "bat" and "back"). A third picture, the distractor, was a semantically related picture. Again, the two patients performed differently: N.G. failed the task in both hemifields, whereas L.B.'s performance was perfect when the pictures were presented in the right visual hemifield. However, L.B.'s performance decreased when the pictures were presented in the left visual hemifield (right hemisphere), though he was still performing at a level above chance. The third task involved matching a written word with a picture. The patients were instructed to associate a word (e.g., "bat") with the picture of a rhyming word (e.g., the picture of a "hat"). Both patients failed the task when the stimuli were presented to the right hemisphere. In the fourth task, the patients had to select from among four written words the two words that rhymed (e.g., "nail" and "male"). Although they were able to perform the control task (i.e., matching the written form with the verbal form in a multiple-choice task), both patients failed the rhyming written words task when the stimuli were presented in the left hemifield (right hemisphere). Therefore, it was logical that they should also fail the fifth task, which required them to associate two rhyming nonwords from an array of four choices presented in the left hemifield. Their failure on the control task, which consisted in matching a written nonword (presented in the left visual field) with its oral form, provided further evidence in support of the right hemisphere's incapacity to perform phonological recoding.

The similar results of these two commissurotomized subjects can be summarized as follows: (1) their right hemisphere was able to associate a word, following written or oral presentation, with its corresponding picture, and to associate a written word with its verbal form, but (2) their right hemisphere was unable to associate two rhyming words or a written nonword with its oral form.

These results demonstrate that the isolated right hemisphere can read and understand words but that it is unable to translate print into sound. These experiments also show that the results of these two subjects cannot be compared. In particular, a question remains concerning the success of L.B. on the homonymy and rhyming tasks with pictures. Indeed, L.B.'s results suggest the possibility of some access to phonological information in pictures but not in reading.

For the purpose of this discussion, it should be kept in mind that phonological recoding is not possible in the isolated right hemisphere. It is possible though to link these findings with those of subjects who have undergone an early left hemidecortication. Indeed, the two patients examined by Dennis, Lovett, and Wiegel-Crump (1981) also experienced difficulties in extracting phonological information from the written code.

Phonological Recoding and the Right Hemisphere of the Normal Subject

As we will see in this section, there are several ways of evaluating the right hemisphere's inability to shift from print to phonology in normal subjects. We will also see that the interpretations of certain studies are questionable and that certain findings are contradictory.

In Moscovitch's study (1976, experiment 3), the task was to determine whether a letter presented in one hemifield rhymed with its oral form presented two seconds earlier. The demonstration of a right hemifield advantage, regardless of the response hand, is supportive of the right hemisphere's inability to convert print into sounds. These results are compatible with those of Zaidel and Peters (1981), although the right-hemifield advantage was not observed for negative answers when the letter and the sound were different.

An experiment reported by Coltheart (1980, 1983), as well as Zaidel and Schweiger (1984), which also supports the absence of phonological decoding by the right hemisphere, is that of Cohen and Freeman (1978). In this study, the effects of pseudohomophony were examined in a lexical decision task. The reaction time for rejecting a nonword was longer when it was homophonic with a word of the language (e.g., kurten) than when it was not (e.g., kolten). The interpretation of this effect is a subject of debate. For some (e.g., Martin, 1982), it may be due to the visual similarity between homophonic nonwords and words of the language. For others (e.g., Patterson & Besner, 1984b), the effect persists even when visual similarity is controlled; in this case, it is thought to reflect the involvement of phonological recoding, which causes a delay in rejecting a nonword that is phonologically similar to a concrete word. Admitting that this interpretation is accurate, the results of the Cohen and Freeman study support the absence of phonological recoding in the right hemisphere. Indeed the effect was observed only for words presented in the right hemifield (left hemisphere). However, this finding has not been met with unanimity. For example, a study conducted by Barry (1981), using the same methodology, revealed a pseudohomophonic effect in both the left and the right hemifields.

Another method is based on an assessment of the regularity effect. This effect reflects the difference between a regular word (e.g., must) and an irregular word (e.g., deaf). In a reading task, reaction times were longer for irregular words than for regular words (Parkin, 1984). Likewise, in a lexical decision task, reaction times were longer for irregular words than for regular words (Parkin & Underwood, 1983; Stanovich & Bauer, 1978). This regularity effect seems to be due to the elaboration of an inappropriate phonological decoding, which, without going into the details of its mechanisms, involves a competition between grapheme–phoneme conversion and lexical information or between the phonological codes of similar words (Glushko, 1979; Henderson, 1984; Kay & Marcel, 1981).

If the right hemisphere is incapable of any phonological recoding, then the regularity effect should not be observed when irregular words are presented in the left hemifield. Parkin and West (1985) asked 16 subjects to name regular and irregular words which were presented unilaterally for 200 msec. The words were matched for frequency and number of letters. The results confirmed the hypothesis: the regularity effect was observed only for words presented in the right hemifield. As for the left-hemifield presentations, a similar number of regular and irregular words were named correctly. Although this finding is indeed supportive of the right hemisphere's incapacity for phonological recoding from print, its limitations should be emphasized. For example, the error rate was high (more than 70%), due to the difficulty of the task (i.e., use of infrequent words and vertical presentations).

Another method is based on the pronounceable or nonpronounceable characteristics of nonwords. Axelrod, Haryadi, and Leiber (1977) found that pronounceable nonwords were read better when they were presented in the right hemifield (left hemisphere) and that there was no difference between the hemifields for nonpronounceable nonwords. However, this advantage was based on a success rate of only about 10%, suggesting a possible floor effect and leading to considerable doubts with respect to the interpretation (Lambert, 1982b).

Using a slightly different method, Young et al. (1984) confirmed these results. They compared laterally presented nonwords that were either pronounceable (e.g., bov) or nonpronounceable (e.g., vbo). They asked 24 normal subjects to name each of the letters of the nonword. Bilateral tachistoscopic presentations, 150 msec in duration, were used. Correct naming was more frequent in the right visual field when the nonwords were pronounceable, but there was no difference between the hemifields when the nonwords were nonpronounceable.

These results suggest that the two hemispheres are able to identify letters, but that the left hemisphere is sensitive to the spelling features of nonwords (whether or not they are pronounceable) while the right hemisphere is not.

In conclusion, if the right hemisphere's incapacity to perform phonological decoding from spelling seems to be well established in commissurotomized patients, it seems that the results from various studies of normal subjects are also supportive of this incapacity of the right hemisphere (Bradshaw & Nettleton, 1983). However, it remains to be determined why several studies (Barry, 1981; Bradshaw & Gates, 1978; Patterson & Besner, 1984a) failed to reach a similar conclusion. Methodological considerations are probably responsible, at least in part, for these contradictory findings.

The Concrete Versus Abstract Meaning of Words

Deep dyslexic patients are much more proficient in reading concrete words, or words with high imagery values, than they are in reading abstract words of similar length or frequency (Coltheart, 1983). One argument to support the role of the right hemisphere in deep dyslexic reading comes from the observation of hemispheric differences in normal subjects with respect to this concrete–abstract dimension. The question of the right hemisphere's potential with respect to the concrete/abstract nature of words was addressed in Chapter 4. It was mentioned that the degree of word concreteness seemed to be one of the factors influencing functional lateralization, with abstract words being lateralized to a greater degree than concrete words. Consequently, this possible capacity of the right hemisphere to process concrete words is compatible with Coltheart's hypothesis.

> It remains reasonable, then, to believe that the right hemisphere can be shown to be superior at dealing with concrete words than at dealing with abstract words. This conclusion allows one to interpret the beneficial effect of concreteness on deep-dyslexic reading within the context of the right-hemisphere account for dyslexia. (Coltheart, 1983, p. 181).

It should be remembered, though, that the evidence from studies of normal and commissurotomized subjects concerns the potential of the right hemisphere, not its actual contribution. Consequently, Coltheart's hypothesis should seek support within the right-brain–damaged population in terms of a specific concrete versus abstract word impairment. As reported in Chapter 4, indirect support for this hypothesis has been provided only by Villardita, Grioli & Quattropani (1987). However, future studies will have to show that right-brain–damaged right-handers exhibit a specific impairment of concrete word processing before it can be used as a premise for a hypothesis which pretends to account for a reading disorder following a left-hemisphere lesion.

Semantic Paralexias and the Right Hemisphere

Coltheart's (1980, 1983) initial hypothesis was that semantic paralexias produced by deep dyslexic patients were due to the contribution of the right hemisphere, whose semantic competence is limited. In reality, the same argument has been used to explain the semantic paralexias produced in any form of aphasia (Goodglass, Graves and Landis 1980; Landis, Graves & Goodglass 1982; Landis, Regard, Graves and Goodglass 1983; Regard & Landis, 1984). It was noted in Chapter 3 that certain authors are skeptical about the possibility of a contribution of the right hemisphere to the residual language of aphasics. It was also noted that there is a lack of adequate means by which to evaluate this contribution. The main objective of the following studies is to produce evidence suggesting a direct contribution of the right hemisphere to the elaboration of semantic paralexias.

Semantic Paralexias and Deep Dyslexia

It should be emphasized that although the presence of semantic paralexias is essential to the semiology of the deep dyslexia syndrome (Coltheart, 1980, 1983), its frequency of occurrence varies from one patient to another. The mechanisms responsible for producing semantic paralexias are numerous (Marshall & Patterson, 1985). However, they can be divided into two main groups: either the semantic system is impaired or word access is difficult (Shallice & Warrington, 1980). These two distinct mechanisms help explain why the actual percentage of semantic paralexias varies from one case to another (Kremin, 1984).

When Coltheart affirmed that semantic paralexias were produced by the right hemisphere, he based his argument on a case described by Gott: "The results of Gott (1973) might be taken as indicating that a propensity to make semantic errors is an intrinsic characteristic of a right hemisphere reading system" (1980, p. 332). The patient described by Gott had undergone a left hemispherectomy when she was 10 years old. Out of a total of 20 words, she could read 9 of them, 5 of which were correct whereas the others produced semantic paralexias (e.g., "book" was read as "poem"). Coltheart admitted that the early age of onset of the patient's difficulties (i.e., 8 years) made it difficult to compare her right hemisphere with that of a dyslexic patient insofar as this hemisphere had probably acquired linguistic capacities that were superior to those of a normal individual's right hemisphere.

The deep dyslexic patient, R.W., examined by Zaidel and Schweiger (1983, 1984) as well as Schweiger et al. (1983), provided more direct evidence. The patient was asked to read aloud 64 words presented in each hemifield. The distribution of the patient's errors was as follows: 63% of the derivational errors (5/8) and 73% of the visual errors (11/19) occurred in the right hemifield (left hemisphere), whereas 73% of the semantic errors (16/22) occurred in the left hemifield (right hemisphere). These results are compatible with the left hemisphere's control of reading (derivational and visual errors) and suggest that semantic errors occur primarily in the left hemifield (right hemisphere). Although the results of this patient were compatible with the hypothesis, it should be noted that paralexias were also found for words presented in the right hemifield. Thus, although the right hemisphere was actually involved in the production of semantic paralexias, its involvement was only partial.

With the exception of this last study, there is little evidence to support an involvement of the right hemisphere in the production of semantic paralexias in deep dyslexia. Moreover, the percentage of paralexias actually observed varies considerably from one study to another. Zaidel and Schweiger (1984) have gone so far as to suggest that there are two subgroups within the deep dyslexia syndrome. Indeed, the involvement of the right hemisphere might exist only in patients such as R.W. who also present deep dysgraphia as well as disorders involving the semantic comprehension of written material. In any case, the patient described by Zaidel and Schweiger provides evidence of a contribution of the right hemisphere but does not provide evidence of the exclusivity of this contribution.

Semantic Paralexias and Aphasia

We next turn our attention to studies that have addressed the same question in unselected populations of aphasic subjects. To evaluate the eventual contribution of the right hemisphere, Goodglass et al. (1980; see also Landis et al., 1982) examined the effects of the concrete/abstract nature of words and of their emotional value on the production of semantic paralexias and paragraphias. It was hypothesized that the involvement of the right hemisphere could be demonstrated if the paralexias occurred primarily for concrete or emotionally charged words.

A group of 32 fluent and nonfluent aphasics were instructed to read and write 36 words, matched for their frequency of use in the written language and divided into three separate categories: concrete words (e.g., boat), abstract words with high emotional value (e.g., fear), and abstract words with low emotional value (e.g., time). Eight subjects made no reading errors at all, whereas only two subjects were unable to read any words at all. The success rates of the other subjects were as follows: 60.6% for the high–emotional value words, 50.8% for the concrete words, and 40.2% for the low–emotional value words. Each factor (i.e., concreteness, emotional value, and frequency of use) had a significant influence on reading performance. An analysis of the semantic paralexias produced by ten of the subjects revealed that they were more frequent for high–emotional value words than for stimuli from the other two word categories. As for writing performance, the success rates of the 18 subjects who were tested followed a pattern similar to that observed for reading performance in five of the subjects. In five of the subjects, the percentage of semantic paralexias was higher for high–emotional value words.

During a second phase of their study, Goodglass et al. (1980) compared the reading performance of the aphasic subjects with that of a group of normals on a laterally presented lexical decision task (Graves, Landis & Goodglass, 1981). The stimuli were identical to those of the previous experiment. A significant correlation was found for words presented in the left hemifield (right hemisphere) but not for words presented in the right hemifield.

According to Goodglass et al. (1980), these results provide two arguments supportive of a contribution of the right hemisphere to the reading performance of aphasics. In reality, it seems that they interpreted in a rather dichotomic manner the two concepts on which their hypothesis was based: (1) the right hemisphere is dominant for the processing of emotional information, and (2) the right hemisphere of normal subjects is better equipped to identify concrete words than abstract words. Again, as was the case in Chapter 4, the reality of these facts is somewhat more complex and one should be cautious. If indeed there is some evidence for the dominance of the right hemisphere for emotional stimuli, this does not necessarily imply a total incompetence of the left hemisphere for the same material (Bear, 1983; Gainotti, 1986; Seron & Van der Linden, 1979). Although the results of a number of studies support the existence of a hemispheric asymmetry on the basis of the concrete/abstract nature of words, several studies have failed to reach a similar conclusion (see Chapter 4).

Nonetheless, the greater frequency of semantic paralexias for high–emotional value words is compatible with the hypothesis of a right-hemisphere contribution, though it is not a very convincing piece of evidence. On the one hand, the number of semantic paralexias is actually quite small (i.e., 10 patients produced a total of only 25 paralexias). On the other hand, the hypothesis implies that the damaged left hemisphere is incapable of producing semantic paralexias, which would be the most logical supposition.

Landis et al. (1983) presented further evidence to support the contribution of the right hemisphere. They examined the relationship between lesion size and the production of semantic paralexias in aphasic patients. Their main finding was that lesion size was greater in patients who produced semantic paralexias than in patients who did not. The authors concluded: "The finding of a connection between size and occurrence of paralexias supports the hypothesis that the emergence of alternative right hemisphere reading processes depends on the relative isolation of this hemisphere" (p. 363). The conclusion that paralexias originate within the right hemisphere is overstated for two reasons. First, it is based on a limited number of errors, which are themselves a subject of debate (Jones & Martin, 1985; Marshall & Patterson, 1983). Second, it postulates that semantic paralexias reflect right-hemisphere activity by assuming that the more massive the lesion of the left hemisphere is, the less inhibition this hemisphere will have on the right. The conclusion is also premature, for it overlooks a more immediate interpretation: semantic paralexias are a direct consequence of left-hemisphere damage (Marshall & Patterson, 1983); the more massive the left-hemisphere lesion is, the less chance there is of the intact areas allowing correct reading, thereby increasing the probability of observing semantic paralexias.

The same group of researchers also found support for this contribution of the right hemisphere from studies of normal subjects. Using very brief tachistoscopic presentations (i.e., 9 msec), Regard and Landis (1984) asked 20 normal subjects to read a series of words ($N = 40$) belonging to four different categories: concrete words with high emotional values (e.g., Blut, blood), concrete words with low emotional values (e.g., Gras, grass), abstract words with high emotional values (e.g., Reue, remorse), and abstract words with low emotional values (e.g., Wert, value). The percentage of errors for the briefly presented stimuli was 75%, of which only 6% was semantic paralexias. Most of the paralexias (50/64) occurred for words that were presented in the left visual hemifield (right hemisphere). This superiority was observed for all four word categories. Regard and Landis concluded that the stimulus presentations were too brief for the left hemisphere to exert its usual control over the right hemisphere and that the right hemisphere was therefore free to produce semantic paralexias. This dynamic conception of interhemispheric relationships will be examined more fully in the next section (Zaidel, 1986a). Although the findings of this study are consistent with those of previous studies of aphasic patients, there are two problems: (1) some of the paralexias which were labeled as semantic paralexias are questionable (Patterson & Besner, 1984b), and (2) the percentage of semantic paralexias produced was very low (i.e., 6% of the total number of incorrect responses). Similar criticism has also been presented by Ellis and Marshall (1978), as well as by Williams and Parkin (1980), concerning a study by Allport (1977) in which normal subjects exposed to brief bilateral presentations produced very few semantic paralexias.

In spite of these limitations, some importance can be given to these various findings, which have the merit of being convergent. The associations between the concrete or emotional nature of words and the lexical decision performance of the right hemisphere in normals, and even their semantic paralexias, are coherent findings in themselves. They are compatible with the right hemisphere's contribution to semantic paralexias in aphasia. However, these findings are indirect; they are not proof of such a contribution and do not justify attributing responsibility exclusively to the right hemisphere. Until further evidence has been provided, one must consider that the damaged left hemisphere is primarily—though maybe not solely—responsible for semantic paralexias (Marshall & Patterson, 1985).

In summary, these findings reveal a convergence of arguments in support of the hypothesis that semantic paralexias are not solely the product of the residual capacities of the left hemisphere, but that they could reflect a possible contribution of the right hemisphere.

Comprehension of Written Words

The right hemisphere's abilities to understand written words were described in Chapter 4. The aim of this section is to report on the right hemisphere's comprehension of written words in deep dyslexia.

To support his hypothesis, Coltheart (1983, p. 176) established a parallel between the semantic comprehension errors of a deep dyslexic (patient G.R., Newcombe & Marshall, 1980) and those of the right hemisphere in two commissurotomized patients (L.B. and N.G.) who were tested by Zaidel (1982) on a word–image association task. In reality, not all deep dyslexics are subject to reading comprehension disorders (Coltheart, et al., 1987). For example, Bishop and Byng (1984) examined the comprehension performance of eight deep dyslexic patients on an 80-item concrete word–picture recognition task. The mean number of semantic errors observed was 3.75. Stated otherwise, a semantic comprehension disorder is not a characteristic of deep dyslexia. Patient G.R. (Newcombe & Marshall, 1980), for example, made 13 errors, whereas patients D.E. and D.W. (Patterson & Besner, 1984a) made no errors at all.

Patterson and Besner (1984a) compared the comprehension of D.E. and D.W. with that of the right hemisphere of patients L.B. and N.G. The logic behind this study was that if the comprehension capacities of deep dyslexics do indeed involve the right hemisphere, then one would expect to find that their performance is similar to what is observed when the right hemisphere of each of the two commissurotomized patients is tested. The tasks included (1) picture recognition of a written word (36 stimuli) with the response being either within or outside of the semantic category (subtests of the BDAE, Goodglass & Kaplan, 1972); (2) picture recognition with four possible choices (PPVT, Dunn, 1965); and (3) completion of written sentences choosing from among four possible responses (WAB, Kertesz, 1982).

On each of the three tasks, the performance of the two deep dyslexic patients was superior to that of the two commissurotomized patients. The comparison is valid only for the patients described here and one must avoid making any generalizations that would apply to all deep dyslexics. It seems logical to attribute the superiority of the deep dyslexic patients to the fact that, unlike commissurotomized patients, they are able to rely on their left hemisphere, even when it is damaged.

An argument against the Patterson and Besner conclusion is that their two deep dyslexic patients showed no signs of any comprehension disorder. It is possible, then, that the involvement of the right hemisphere is to be found only in deep dyslexic patients with comprehension disorders. Thus the deep dyslexic patient R.W. (Zaidel & Schweiger, 1983) produced semantic errors on a task that required the comprehension of written words by way of picture recognition, much like the right hemisphere of L.B. and N.G. Although the performance of this patient on a lexical decision task was comparable to that of the two split-brain patients, once again the results should not be extrapolated to other deep dyslexics with similar semiology (Zaidel & Schweiger, 1983).

In summary, it is best to say that deep dyslexics understand with both their hemispheres, that the involvement of the right hemisphere is possible, and that it varies among the patients who are grouped under this syndrome.

> Thus the right hemisphere may contribute to fixed specific processing modules missing from the left hemisphere, or it may be called upon to help the left hemisphere when the latter is overloaded or fatigued, and therefore nonfunctional. In short, the correspondence between deep dyslexia and right-hemisphere reading is at best one of partial identity. (Zaidel & Schweiger, 1984)

The Problem of Pure Alexia

The major argument against the right hemisphere's role in the comprehension of writing in deep dyslexia lies in the absence of comprehension in patients with pure alexia. All of the hypotheses concerning the relationships between the two hemispheres for reading (Landis, Regard et al., 1980; Patterson & Besner, 1984a; Zaidel, 1986a; Zaidel & Schweiger, 1984) should take pure alexia into account. The observation that patients with pure alexia have poor comprehension abilities (Kremin, 1984), in spite of having an intact right hemisphere, led Coltheart (1983) to reexamine his earlier hypothesis (1980).

We begin by briefly describing the major semiological characteristics of alexia without agraphia (pure alexia, agnosic alexia, letter-by-letter alexia) before describing how Coltheart (1983) attempted to reconcile the right-hemisphere hypothesis in deep dyslexia with the existence of alexia without agraphia. According to Alajouanine, Lhermitte, and Ducarne de Ribaucourt (1960), reading words is much more affected than reading letters in alexia without agraphia: the patient attempts to read the word, letter by letter. Spontaneous writing and dictation are normal, but the patient is unable to read what he has written. The naming of spelled words is normal. There is little or no disturbance of oral expression or comprehension. The lesions responsible for alexia without agraphia include a left occipital lesion, most often an ischemia of the posterior cerebral territory, resulting in a right lateral hemianopia. It is often associated with a lesion of the callosal pathways, that is, fibers of the splenium or intraoccipital callosal fibers (Damasio & Damasio, 1983, 1986; Déjerine, 1892; Geschwind, 1965).

The question raised by the alexia without agraphia syndrome is quite simple. If, as Coltheart (1980, 1983) suggested, the right hemisphere possesses the capacity to access the comprehension of written words, in particular concrete words, pure alexic patients should be able to use this capacity and therefore understand words which they are not able to read aloud. In reality, the writing comprehension of alexics is altered in parallel with their difficulty in reading aloud; comprehension is possible only after identification of each of the letters has been completed (see, for example, Warrington & Shallice, 1980). The two tasks used by Patterson and Kay (1982) illustrate this comprehension disorder. Following the presentation of a written word, subjects were asked either to make a semantic judgment or to point to a picture belonging to the same semantic category. Then they were asked to read the word out loud. As it turned out, the judgments and choices for words that were read incorrectly to begin with were incorrect themselves. These patients do indeed show a written-word comprehension deficit. To explain this finding, contradictory with his earlier hypothesis, Coltheart was forced to propose a more complex hypothesis.

According to Coltheart (1983), the right hemisphere is capable of processing certain written words (especially concrete words) and then transferring this processed information ("categorized," in Coltheart's term) to the left hemisphere. Two types of categorization are likely to be carried out by the right hemisphere: (1) orthographic categorization, which consists in the decomposition of a word into its individual letters, and (2) semantic categorization. Finally, it is possible that some information is transferred from the right hemisphere to the left hemisphere without undergoing any processing at all (i.e., the right hemisphere is a simple relay).

Coltheart assumes that in the case of deep dyslexia, the only information that can be processed by the left hemisphere is the semantic information coming from the right hemisphere. The left hemisphere is therefore dependent on the quality of the right hemisphere's categorization, which is better for concrete words. To explain alexia without agraphia, it would have to be assumed, however, that semantic information processed by the right hemisphere cannot be transferred to the left hemisphere. Only information that has been directly relayed by the right hemisphere is received by the left, which then proceeds to a letter-by-letter decoding. Stated otherwise, Coltheart added the nontransfer of a particular type of information (i.e., semantic categorization performed by the right hemisphere) to the callosal disconnection hypothesis put forward by others (Geschwind, 1965) in order to explain alexia without agraphia. These explanations are ad hoc only. It is difficult to verify, as Coltheart suggested, any sort of relationship between an anatomical disconnection of the splenium (which is far from being constant; see Damasio & Damasio, 1983) and a particular type of information. It should be pointed out that Sidtis et al. (1981) described a patient in whom the transfer of semantic information remained possible even after surgical sectioning of the splenium. Indeed, the patient was able to describe the characteristics of words presented in his left visual field. To explain alexia without agraphia and still maintain his right-hemisphere hypothesis of deep dyslexia, Coltheart introduced the concept of the inhibition exerted by the damaged left hemisphere on the right. However, he reported no evidence of this inhibition as being specific to a particular type of information. The same concept had already been used to explain the absence of recovery from pure alexia due to ischemic causes (Hécaen, 1976; Jeannerod & Hécaen, 1979). However, the term "inhibition" itself hides our ignorance; we have yet to find methods that are adequate to test the mechanisms supposedly governing this inhibition (Bradshaw & Nettleton, 1983).

According to Zaidel et al. (1986a) and Landis, Regard, et al. (1980), this inhibition should not be considered as permanent but rather as dependent on the nature of the task to be performed. This dynamic conception is based on a study conducted by Landis, Regard, et al. (1980). It concerned a patient with pure alexia secondary to a left occipital glioblastoma. When words were presented in his left visual field, using a period of presentation that was short enough to prevent letter identification, the patient was able to correctly point to corresponding objects. Conversely, when the improvement of the patient's condition enabled him to name briefly presented letters, then he was no longer able to exhibit right-hemisphere comprehension. This observation is to be viewed with caution, considering the turmoral etiology and the few words tested (seven words). However, its theoretical importance makes it worthy of mention. The conception is the same as that previously suggested in a study of the production of semantic paralexias in normals (Regard & Landis, 1984). According to Zaidel (1986a), reading involves an interaction between the two hemispheres under the control of a regulating center, possibly located in the left hemisphere. Following a lesion of the left hemisphere, the impairment of this regulating mechanism does not seem to eliminate the inhibition which is normally exerted on the intact and competent right hemisphere (Zaidel, 1986a). This inhibition could be eliminated if the task were beyond the domain of competence of the left hemisphere yet still accessible to the right hemisphere.

In summary, it is apparent that Coltheart's hypothesis does not fit well with the absence of reading comprehension in pure alexia. The presence of a right-hemisphere potential for comprehension is not enough to explain the comprehension of deep dyslexics and the semantic errors of some of them. The concept of inhibition defended by Coltheart (1983), put forward to reconcile the semiology of pure alexia with that of deep dyslexia, has the inconvenience of all the explanations which incorporate it. There is no way of evaluating this hypothesis adequately. Studies similar to that of Landis, Regard, et al. (1980) with alexics should be able to determine if it is indeed possible to find the stimulus conditions that would eliminate the left hemisphere's inhibitory effect on the right hemisphere during reading. We know from Hécaen's (1976) follow-up studies of left occipital lobectomized patients that recovery from pure alexia is favorable, which suggests that, freed from the control of the left hemisphere, the right hemisphere is capable of contributing to reading.

Conclusion

The interest of Coltheart's (1980, 1983) hypothesis lies in its discussion of the existence of a standard equipment of the right hemisphere for reading: an absence of phonological recoding, a lexicon that is essentially concrete, and a capacity for semantic categorization. We have examined, in this chapter as well as in the preceding chapter, the evidence from a number of studies supporting such a potential of the right hemisphere. However, the experimental evidence does not allow one to conclude that this right-hemisphere potential is involved in cases of left-hemisphere lesions or in the reading behavior of normal subjects. This chapter, which has focused on deep dyslexia, illustrates how important it is not to merge these two conceptions.

The semiology of deep dyslexia has so many analogies with this supposed potential of the right hemisphere that Coltheart (1983) went one step further to attribute this semiology to the actual contribution of the right hemisphere. In reality, an examination of the different implications of this hypothesis leads to the supposition of only a partial contribution of the right hemisphere. By the same token, although the right hemisphere is involved in the elaboration of semantic paralexias, it should not be considered as solely responsible. On the basis of the findings presented in this chapter, one is forced to admit that our knowledge concerning the potential contribution of the right hemisphere is greater than that concerning the relationships between the two hemispheres or the nature of the mechanisms underlying this right-hemisphere contribution. We do not have enough data to adequately define the concepts of competition and inhibition which remain at the heart of the conjecture. Is the left hemisphere's inhibitory effect over the right a permanent one or must we consider, like Zaidel (1986a), that there is a constant integration of the processing carried out by each hemisphere? There is a finality to this conception: the results of such processing are either congruent or discordant. There is a facilitation of transfer when the integration of processing within each hemisphere is required and an inhibition of transfer when such processing is discordant, in which case it is the decision of the left hemisphere that prevails.

Our knowledge of the right hemisphere's potential for written words could result in its being attributed certain of the processes involved in reading: its capacity to process written words in a lexical manner, its capacity to perform this proessing for rapid presentations, and its lexicosemantic capacities all lead to the assumption that the right hemisphere is useful for the recognition of word patterns, a necessity for rapid and efficient reading. Although this conception is plausible, it remains to be demonstrated.

6
Prosodic Aspects of Speech

The interest of aphasiologists in the prosodic aspects of language is obvious in many aphasiological descriptions. Classic clinical teaching emphasizes the difference between reduced verbal expression and preserved intonation, which, for example, can help aphasics express their helplessness or can underlie the automatic production of songs and prayers.

> The speechless patient may utter "yes" or "no" or both in different tones. . . . It is not a proposition but an interjection of a mere vehicle for variation of voice expressive of feeling. (Hughlings-Jackson, 1879, p. 210)

> In many cases of motor aphasia, the faculty of singing words is conserved in spite of complete inability to speak a single word. In such cases, the patient probably sings by means of the right hemisphere. (Henschen, 1926, p. 118)

The notion of prosody was first introduced by Monrad-Krohn (1947) in reference to the verbal behavior of certain brain-damaged patients. He distinguished four different types of prosody:

1. *Intrinsic prosody* refers to the intonation contours that distinguish a declarative from an interrogative sentence.
2. *Intellectual prosody* refers to the placement of stress, which gives a sentence its particular meaning. For example, if the sentence "He is clever" is stressed on "clever," it has a sarcastic connotation, whereas if it is stressed on "he," it means that everyone else is not.
3. *Emotional prosody* conveys joy and anger, as well as other emotions.
4. *Inarticulate prosody* consists of grunts or sighs and conveys approval or hesitation.

Monrad-Krohn also described three disorders of prosody that occur subsequent to a lesion of the nervous system, though he did not consider lesions of the right hemisphere.

1. *Hyperprosodia* is observed in manic states and in motor aphasia.
2. *Dysprosodia*, also qualified as ataxic, is characterized by a foreign pseudo-accent[1] and was first observed in a patient recovering from Broca's aphasia. Dysprosodia corresponds to preserved, albeit abnormal prosody.

3. *Aprosodia* is the inability to produce any variation at all in prosody, such as in the case of parkinsonian patients.

The distinction between linguistic prosody (i.e., intrinsic and intellectual) and emotional prosody has been the object of much of the current research. This distinction overlaps, at least in part, the suggestion made a number of years earlier by Hughlings-Jackson (1879) in his distinction between propositional language and emotional language. The relationship between emotional prosody and the right hemisphere also underlies all of the descriptions of anosognosia in patients with lesions of the minor hemisphere (Gainotti, 1972; Weinstein & Kahn, 1955). The behavior of these patients and the lack of any apparent affectivity in their speech are seen as indicators of an indifference and a lack of expression, which is in sharp contrast with the catastrophic reaction of aphasics (Goldstein, 1948).

Though the relationship between the right hemisphere and emotional prosody was implicit for a long time, almost two decades went by before any specific studies of this relationship were conducted. The reference to the preserved prosodic capabilities of aphasics and the privileged relationship between the right hemisphere and emotional behavior were to have an impact on a number of studies and help explain why some authors attempted to describe the mechanisms of prosody in a dichotomous manner, right brain versus left brain. Likewise, the effectiveness of Melodic Intonation Therapy (Helm-Estabrooks, 1983) with certain aphasics should not be considered in itself as sufficient evidence of the presumed intact right hemisphere's ability to take over all aspects of prosodic functioning. In this section, we would like to insist on the difference between the complexity of what is covered by the term prosody, the small number of models used to describe its various aspects, and the gross dichotomy between linguistic prosody and emotional prosody underlying many of the studies of right-brain–damaged patients. This dichotomous conception of prosody should not be considered as an accepted fact; it is merely a hypothesis. A priori, there is nothing to suggest that this conception is pertinent to the understanding of the mechanisms underlying prosodic functioning. For example, there are no acoustic parameters enabling one to differentiate emotional from linguistic stimuli. Moreover, on the basis of this hypothesis, most studies have attempted to establish a relationship between the presence of prosodic disorders and the hemispheric localization, right or left, of the lesion. The superiority of the left hemisphere for language processing is a well-established fact, and it would be logical to assume that linguistic prosody is also controlled by the left hemisphere. Conversely, the possible superiority of the right hemisphere in controlling the emotional aspects of behavior (Bear, 1983; Gainotti, 1972) would suggest that the same is true for emotional prosody. As we will see, these caricatured conceptions are too simplistic and warrant the consideration of alternative hypotheses.

---

[1]Cf. the phonetic disintegration syndrome (Alajouanine, Ombredane, & Durand, 1939).

Definitions

Before reviewing the various studies, it is important to provide definitions for some of the terms that will be used with respect to emotional and linguistic prosody. The two aspects of linguistic prosody most studied are stress and intonation. Stress can be defined as the emphasis placed on a syllable or on a group of phonemes. Two types of stress have been described in the literature: (1) lexical stress is used to distinguish two words with the same phonemic structure (e.g., "redcoat," a British soldier, versus "red coat," a piece of clothing), and (2) emphatic stress is used to convey a different idea depending on the word's position in the sentence (e.g., in the sentence "*John* drives the car" it is surprising that John is the driver, whereas in "John *drives* the car" it is the fact that he actually drives the car which is emphasized). Emphatic stress corresponds to what Monrad-Krohn (1947) called *intellectual prosody.* The intonation contour, for its part, concerns the whole sentence and translates its modality (e.g., declarative, interrogative, imperative, or exclamatory). Emotional prosody is marked by intonation. Depending on the intonation, the expressed emotion denotes joy, sadness, anger, surprise, or any other emotional state that can be conveyed by oral language.

To fully understand the studies that follow, a brief description of the parameters of the acoustic signal that contribute to prosody will be helpful. The three major components of the acoustic signal are its fundamental frequency (F_0), its duration, and its intensity. The fundamental frequency of the sound wave varies as a function of time. The highest values are referred to as peaks; the lowest values are referred to as valleys. The difference between these two extreme values is called the range. Thus the corollary of sentence intonation is the variation of the fundamental frequency during the emission of a sentence. For example, a declarative sentence usually terminates with a drop in intonation, and therefore in the F_0, whereas an interrogative sentence usually terminates with a rise in intonation. Happy sentences are usually emitted with higher intonation (F_0), with larger ranges, and with more variability than sad sentences (Cosmides, 1983; Williams & Stevens, 1972).

Emotional Prosody

The objective of this section is to examine the difficulties right-brain–damaged patients have with respect to the production and comprehension of emotional prosody. The issues to be examined include the nature of the impairment, its dubious restriction to the emotional component of prosody, and its specificity to right-brain–damaged patients.

For a number of reasons, we begin this section with a review of some of the experiments conducted by Elliot Ross and his colleagues. One reason for doing so is that these researchers were instrumental in attracting the attention of clinicians to the systematic study of prosody in right-brain–damaged patients. Indeed, the observations of Ross and his colleagues contain descriptions of the spontaneous prosodic behavior of right-brain–damaged patients.

The first case described by Ross and Mesulam (1979) was that of a teacher hospitalized for a sensorimotor disorder of the left side of the body following an ischemia of the inferior portion of the ascending frontal and parietal convolutions. Upon resuming her activities, one month after her ictus, the teacher complained of her weak monotonous voice. She could not modulate her voice to express her anger, her frustration, or her disapproval. In emotional situations, she was unable to adequately express how she felt and her tears seemed unnatural and superficial to her relatives. On the other hand, the patient affirmed that she understood the meaning of the intonation and gestures of her family. Within eight months, this symptomatology disappeared and the patient recovered a normal use of intonation in her expression.

The second case was that of a man, the victim of a partially regressive disorder of the left side of the body, and presenting a disorder of prosody. This disorder persisted five years later. His voice was monotonous and, in spite of his efforts, could not convey the warm or jocular intonation that he wanted to use. The patient had no difficulty understanding the emotions expressed by his family. A CT scan revealed an ischemia of the ascending frontal and parietal convolutions extending into the subcortical structures. Using the terminology proposed by Monrad-Krohn (1947), Ross and Mesulam (1979) defined this impossibility to modulate emotional prosody as "aprosodia." The lesion localization common to both these patients—the ascending frontal and parietal convolutions—led the authors to speculate that the anterior portion of the right hemisphere was involved in the expression of prosody. Using the argument of the poor comprehension of emotional prosody that was observed in Heilman's (1975) patients, whose lesions were temporoparietal, Ross and Mesulam hypothesized an anteroposterior partitioning within the right hemisphere for the processing of prosody, a mirror image of the left hemisphere's partitioning for language. They also suggested that the subcortical lesion in the second patient was responsible for the permanent nature of his aprosodia.

More recently, Ross (1981) provided further evidence for this mirror-image conception of the right hemisphere in a report of 18 additional cases, although only 10 were described in detail. The examination protocol was similar to the one used in the clinical assessment of aphasia. The assessment of prosodic expression was based on the quality of intonation in responses to questions of an emotional nature (e.g., questions concerning the illness, questions concerning emotionally laden situations). Repetition of prosody was assessed by how well the patient imitated the intonation of a sentence spoken by the examiner (e.g., happy, sad). The comprehension of prosody was evaluated by asking the patient to describe the intonation of the examiner, who was positioned behind him to avoid any decoding by way of gestural clues.

Ross (1981) also provided an anatomoclinical description for each of the ten cases, using a nosology similar to the one used for language disorders. Three of the patients were described as presenting motor aprosodia because of the association of aprosodic expression, poor repetition, and good comprehension. All three patients had infarcts of the right frontoparietal region. The association of normal prosodic expression with disordered repetition and comprehension characterized the so-called sensory aprosodia of one patient with an ischemic lesion affecting the right posterior temporal region as well as the inferior parietal lobule. Another patient was described as having global aprosodia following a massive right-sylvian infarct. Two patients demonstrated what was called transcortical motor aprosodia, that is, the association of aprosodic expression with normal repetition and comprehension. In one of the patients, the lesion was a massive tumor of the right frontal, parietal, and temporal lobes. In the other patient, the infarct involved the anterior arm of the internal capsule, the putamen, and the head of the caudate nucleus. One patient was labeled transcortical sensory aprosodia, since expression and repetition were normal but comprehension was disrupted. The lesion was a right intratemporal hematoma. Mixed transcortical aprosodia—normal repetition with poor comprehension and expression—was observed in one patient. The CT scan showed a massive ischemia in the right frontoparietal region and a smaller infarct within the left frontal region. Finally, one patient was described as having motor aprosodia with pure prosodic deafness. An ischemic lesion affecting the right frontal operculum, the insula, and the medial portion of the third temporal convolution was believed responsible for this disorder.

The conclusions of this study were overly optimistic: "This dominant language contribution by the right hemisphere seems to be anatomically and functionally organized along the lines of propositional language in the left hemisphere" (Ross, Harney, de Lacoste-Utamsing, & Purdy, 1981, p. 745). These conceptions, which were reaffirmed in subsequent papers (Gorelick & Ross, 1987; Ross, 1984a,b), are doubtful for several reasons:

1. No validation of the examination protocol was conducted. There was only one judge, the examiner himself. Previous studies (e.g., Dordain, Degos, & Dordain, 1971) warned of a lack of concordance between clinical judgments and acoustic analysis of speech. In addition, more recent studies (e.g., Danly & Shapiro, 1982) revealed a difference between the clinical impression of the monotonous voice of Broca's aphasics and acoustic measurements which suggest, on the contrary, a more important variation in the fundamental frequency. Finally, experiments conducted by Tompkins and Flowers (1985) revealed a difficulty in obtaining subjects' agreement as to the emotional nature of the intonation heard.
2. A parallel with the anatomoclinical classification of aphasia, itself a source of debate (e.g., Basso et al., 1985; Poeck et al., 1984), has no theoretical justification. In any case, the number of patients involved in these studies is far too small for a taxonomic study. Furthermore, Ross et al. (1981) themselves described a case of motor aprosodia in which the lesion site was not compatible with the proposed taxonomy. Although the symptomatology corresponded to the criteria of motor aprosodia, the results of the anatomopathological examination revealed a lesion of the right capsule.

3. Prosodic disorders (Ross, 1981) are most frequent during the first three days following an ictus and are unstable, thereby suggesting that reactional emotions could be involved in the disorder, without being attributed to any specific localization.
4. The assertion that there are no negative cases—"all patients who I examined who had a right peri-Sylvian lesion also had an aprosodia" (Ross, 1981, p. 561)—is unjustified (see, for example, the counterexample of Lebrun, Lessinnes, De Vresse, & Leleux, 1985). First, the author failed to provide any evidence concerning the eventual difficulties of left-brain–damaged patients with the same protocol. Second, disorders of the comprehension of emotional prosody have also been described in aphasics (e.g., Schlanger, Schlanger, & Gertsman, 1976). The studies which follow, and which are less ambitious than those of Ross and his co-workers, demonstrate that the assessment of emotional prosody is not as simple as Ross (1984a,b) suggested.

Expression of Emotional Prosody

The flat, monotonous speech in the patients described by Ross and Mesulam (1979) has also been reported in other studies, though in most of them prosody was not the main focus of attention. This characteristic has either been mentioned as being indicative of indifferent affective behavior (Gainotti, 1972), or it has been reported during the course of patients' responses to various tasks (e.g., Gardner et al., 1983; Joanette, Lecours, et al., 1983; Marcie et al., 1965). For methodological reasons, it is difficult to design a systematic study of spontaneous emotional prosody. This helps explain why the only two studies to have focused specifically on prosodic production (Shapiro & Danly, 1985; Tucker, Watson, & Heilman, 1977) have little to do with the assessment of spontaneous prosody. As we will see, the tasks used in these two studies were very demanding.

Tucker et al. (1977) examined the production of emotional intonation in eight right-brain–damaged patients and eight neurologically intact patients. Following the presentation of a neutral sentence, the subjects were instructed to repeat the sentence with a particular intonation as specified by a single word (e.g., anger, joy, sadness, or indifference). The productions of the patients were recorded and played back to three judges, who were asked to determine the intonation of each sentence using one of the four words indicated above. The results revealed a significant impairment of the right-brain–damaged patients. They were unable to follow the instructions that required them, albeit artificially, to assign a particular emotional intonation to a sentence. The only instruction for which their scores were similar to those of the normal patients was the one calling for an expression of indifference. Since we know that the eight right-brain–damaged patients all presented a neglect syndrome, and that their behavior was described as indifferent, the results obtained by Tucker et al. may have simply reflected the global indifference of these patients.

Shapiro and Danly (1985) conducted an acoustic analysis of emotional productions obtained during a reading task. An introductory paragraph outlined a context which induced the subjects to read a target sentence with either a happy or a sad intonation. The performance of 11 right-brain–damaged patients was compared with that of 5 aphasics and 5 normals. The acoustic translation of the happy or sad intonation was observed in all three groups of subjects. Happy sentences were read with a higher fundamental frequency, a larger range, and a greater variability than sad sentences. The results also showed that right-brain–damaged patients with anterior (*N* = 3) or central (*N* = 3) lesions were more restricted in their modulation of the fundamental frequency than either subjects with posterior lesions (*N* = 5) or aphasics and normals. However, one should not jump to the conclusion that prosodic expression of emotion varies as a function of the anteroposterior site of the right-hemisphere lesion. Indeed, the analysis of individual results revealed a great amount of intersubject diversity in performance. As it stands, the results of this experiment suggest that (1) the acoustic parameters that differentiate a happy intonation from a sad one appear to be preserved in right-brain–damaged patients; (2) there are certain right-brain–damaged patients whose acoustic analysis is abnormal, and (3) there is no anatomical evidence adequate to differentiate between right-brain–damaged patients in terms of their expression of emotional prosody.

Comprehension of Emotional Prosody

To evaluate the comprehension of emotional prosody, Heilman, Scholes, and Watson (1975) instructed their subjects to indicate which one of four faces expressed the emotion conveyed by the intonation of a sentence. The semantic content of the sentences was neutral. However, they were produced with an emotional intonation that expressed anger, joy, sadness, or indifference. Right-brain–damaged subjects (*N* = 6) made significantly more errors than aphasic subjects (*N* = 6), to the extent that their performance was at the level of chance. Tucker et al. (1977) evaluated the capacity of right-brain–damaged subjects to name the emotional intonation of a sentence. Once again, performance was significantly poorer in right-brain–damaged subjects (*N* = 11) than in aphasics (*N* = 7), and was at the level of chance.

Contrary to the preceding studies, Schlanger et al. (1976) and Seron, Van Der Kaa, Van Der Linden, and Remits (1982) compared the performance of aphasics with that of normal subjects. Using a facial recognition task similar to the one used by Heilman et al. (1975), Schlanger et al. found no difference between right-brain–damaged subjects (*N* = 20) and aphasics (*N* = 40), who were divided into two equal groups on the basis of the severity of their aphasia. The emotions examined were anger, joy, and sadness. The rate of success was similar to the right-brain–damaged group (70%) and each of the two groups of aphasics (severe, 67%; nonsevere, 77%), with the score of normals being 93% (Schlanger, 1973). The distribution of the errors as a function of the emotion type was similar in all three groups. Furthermore, the sentence content (i.e., whether it was semantically neutral or meaningless) had no influence on performance. Seron, Van Der Kaa, Van Der Linden, and Remits (1982) later confirmed the failure of aphasics on a task similar to the one used by Schlanger et al. (1976), but which also included sentences with a strong emotional content that was either congruent or not with the intonation (e.g., "My cat is dead" with a sad intonation; "I won the big prize" with a sad intonation). Regardless of the type of sentence, aphasics (i.e., 14 Wernicke's, 9 Broca's, and 4 global) made more errors than normal subjects. The results were correlated with the severity of the oral comprehension deficit but not with the aphasia type. In summary, the results of these two studies tend to suggest that when intonation is associated with a verbal message, its emotional content is not well understood by aphasic subjects.

Two explanations can be offered for the difference between, on the one hand, the results of the Heilman et al. (1975) facial recognition task and the Tucker et al. (1977) naming task, which revealed a greater disorder of the comprehension of emotional prosody following a lesion of the right hemisphere, and, on the other hand, the results of teh Schlanger et al. (1976) facial recognition task, which revealed a similar deficit regardless of the hemisphere affected. The first explanation concerns the particular nature of the right-brain–damaged subjects. More subjects participated in the Schlanger et al. (1976) study, suggesting greater diversity in their lesion sites, whereas the subjects in the Heilman et al. (1975) and Tucker et al. (1977) studies shared the particularity of having a parietal lesion and presenting hemineglect. It is likely that these patients were more disposed to disorders involving the comprehension of emotion. The second explanation concerns the selection of the aphasic subjects. Indeed, Heilman et al. (1975) and Tucker et al. (1977) selected patients with conduction aphasia, whereas several different types of aphasia were taken into consideration in the Schlanger et al. (1976) and Seron et al. (1982) studies.

Two comments concerning the methodology of these studies are also warranted. First, the intonation used in the various studies is somewhat artificial. The stimuli were constructed by asking normal subjects to exaggerate a given intonation. This is not intonation as it might be produced under normal circumstances. Second, the choice of facial expressions of emotion as a mode of response (Heilman et al., 1975; Tucker et al., 1977) is questionable, since facial expressions in themselves are likely to cause problems for right-brain–damaged subjects (Bruyer, 1983).[2]

To specify the exact nature of the prosodic comprehension disorder, Tucker et al. (1977) conducted a somewhat different experiment. Whereas previous studies had involved an emotion-identification task, this study involved an emotion-discrimination task. The subjects had to decide if two sentences that were either similar or dissimilar in their emotional intonation had the same verbal content. Once again, right-brain–damaged subjects ($N = 11$) were unable to perform adequately (chance level) and were significantly worse than conduction aphasics ($N = 7$). The failure of right-brain–damaged subjects on this

[2]While on this subject, we would like to draw attention to a study by Benowitz et al. (1983) in which it was found that right-brain–damaged subjects had greater difficulty understanding emotional intonation and facial expressions, though there did not seem to be any correlation between these two emotional modalities.

task warrants two remarks. First, their failure to discriminate between the two sentences is indicative of a disorder at an early stage of auditory perception. This suggests that disorders of a similar nature might also exist for other prosodic stimuli regardless of their function. In a discrimination task, two stimuli which share the same emotional prosody also share a certain number of acoustic features. It is therefore possible that the failure of the right-brain–damaged subjects and the success of the conduction aphasics are more the consequence of acoustic processing itself than of emotional categorization. Second, since these right-brain–damaged patients are described as indifferent and fail all of the prosodic tasks they are given (both in comprehension and in production), shouldn't it be the same for a task of a different nature? In fact, it could be that the impairment of right-brain–damaged subjects reflects a more general attitude of indifference rather than a specific impairment. If this were the case, then emotional prosody deficits would be only one example of this new attitude.

In another study, Heilman, Bowers, Speedie, and Coslett (1984) sought to test comprehension when only emotional prosody was involved, in other words, when phonetic, semantic, and syntaxic components had been eliminated. In this study, sentences containing one of three possible intonations (joy, anger, or sadness) were filtered, thus making their verbal content unintelligible. The subjects had to either name the intonation heard or point to the corresponding facial expression in an array of three possible choices. Right-brain–damaged subjects ($N = 8$) and aphasics ($N = 9$) did not perform as well as normals (88.2% correct). However, the score of the right-brain–damaged group (52.5%) was significantly lower than that of the aphasic group (78.9%). It should also be noted that neither the positive (i.e., joy) nor the negative nature (i.e., anger, sadness) of the intonation had an effect on the choices of these two groups of subjects, which is in agreement with the findings of Seron et al. (1982) in a population of aphasics.

A descriptive summary of the evidence presented thus far would say that any cerebral lesion causes more or less difficulty with the decoding of emotional prosodic intonation, whether or not it is associated with verbal information. Beyond this description, one has to try to understand the nature of the contribution of each hemisphere. The questions could be stated as follows: First, if the difficulty of processing emotional prosody is present at an early stage of perception (Tucker et al., 1977), to what extent is this perceptual disorder related to the emotional nature of prosody? To answer this question, one would have to study the discrimination of emotionally neutral prosodic stimuli to see if the difficulties with these stimuli are similar to the difficulties observed with linguistically determined prosody. Second, considering the implication of the left hemisphere (Heilman et al., 1984; Schlanger et al., 1976; Seron et al., 1982) for which aspects and beginning at which level of complexity is it involved?

These questions were examined by Tompkins and Flowers (1985), who studied 11 right-brain–damaged subjects (including 6 with hemineglect), 11 left-brain–

damaged subjects (including 10 aphasics), and 11 normal subjects.[3] In an attempt to answer the first of the two questions outlined above, Tomkins and Flowers examined the discrimination of filtered sentence pairs (50 pairs) containing either a neutral or an emotional intonation. The group of right-brain–damaged subjects made significantly more errors than the other two groups of subjects who performed similarly. The discrimination errors involved emotionally neutral stimuli as well as emotionally stressed stimuli. Furthermore, the tonal memory scores (Seashore, Lewis, & Saetveit, (1960) of the right-brain–damaged subjects were inferior to those of the other subjects. There are two arguments here to suggest that part of the difficulty of right-brain–damaged subjects resides not in emotional prosody per se but, more fundamentally, in the perceptual decoding of prosodic information.

Tompkins and Flowers (1985) also examined the question concerning the nature of the cognitive operations involved. Two emotion-identification tasks, consisting of semantically neutral sentences ($N = 40$), were used. The subjects were instructed to choose a word that described the emotion conveyed by the sentence (e.g., anger, joy). The only difference was that in one task the subject's choice was limited to two words, whereas in the other there were four possible choices. When the choice was limited to two words, right-brain–damaged subjects made significantly more errors than aphasics or normals, who performed at about the same level. However, when four choices were provided, right- and left-brain–damaged subjects demonstrated a similar error rate, which was higher than the one observed in the normal subjects.

Although their errors were quantitatively similar, there are two arguments to suggest a different mechanism in the right- and left-brain–damaged subjects. The first concerns the distribution of the errors (i.e., the nature of the emotions identified incorrectly), which was similar in normal and left-brain–damaged subjects but not in right-brain–damaged subjects. The second is the presence of a correlation, in the group of right-brain–damaged subjects, between the results on the discrimination and identification tasks, whereas no such correlation was found in the other two groups. This correlation provides further evidence to the effect that the major difficulty of right-brain–damaged subjects lies in the extraction of prosodic information, whereas the difficulty of left-brain–damaged subjects lies in the complexity of the task. The involvement of the left hemisphere is all the more important if the task requires integrating operations from other sources (e.g.,

---

[3]The Tompkins and Flowers study is also interesting in light of the technique they developed to generate prosodic stimuli devoid of the artificiality of the intonations used in other studies. The intonations were recorded during the normal reading of a text by control subjects, and later played back to five judges. From an initial sample of 192 intonations, only 64 received a unanimous judgment. This illustrates the difficulty in obtaining agreement, even among normal subjects, with respect to the criterion of emotional prosody, and therefore casts doubt over a number of studies in which the sole judgment of the examiner sufficed to describe the prosody of a patient (e.g., Ross, 1981).

words, facial expressions). Thus the Tompkins and Flowers (1985) study confirmed the existence of a disorder of emotional prosody in right-brain–damaged subjects but also provided some arguments suggesting that this disorder is not attributable to the emotional nature of prosody. As for the difficulties of the aphasic subjects, these can best be explained by the involvement of the left hemisphere in any complex task (Feyereisen & Seron, 1982a).

Bowers, Coslett, Baven, Speedie, and Heilman (1987) offered two interpretations of the right-brain–damaged patient's impaired comprehension of emotional prosody. The first interpretation (the so-called processing-defect hypothesis) is that a lesion of the right hemisphere could result in a global misperception or a miscategorization of emotional intonation. According to this interpretation, there is no contribution of left-hemisphere mechanisms to the understanding of emotional intonations. Thus the comprehension of emotion by right-brain–damaged subjects should not be modified by any changes in the semantic content of emotionally intoned sentences. The alternative interpretation is that the left hemisphere may have some capacity to mediate emotional prosody. In the event of a right-brain lesion, Bowers et al. hypothesize that the left hemisphere may be distracted (the so-called distraction-defect hypothesis) and attend only to the semantic content rather than to the emotional prosody. Accordingly, the comprehension of emotional prosody should be affected by any changes in the semantic content. Three groups of subjects (nine right-brain–damaged, eight left-brain–damaged, and eight controls) listened to emotionally intoned sentences. They were required to identify the intonation by pointing to the corresponding face from among four choices. On half of the trials, the emotional prosody was congruent with the semantic content of the sentence; on the other half it was incongruent. The right-brain–damaged subjects performed significantly worse than either the left-brain–damaged or the control subjects, which did not differ from one another in either the congruent or incongruent condition. Both groups of brain-damaged subjects performed poorly in the incongruent condition, but the right-brain–damaged subjects more so. The incongruent trials were then analyzed according to the extent of the discrepant information conveyed by the semantic and prosodic messages. It was found that the impairment of the right-brain–damaged subjects varied with the degree of the discrepancy between semantic and prosodic information. Conversely, performances of left-brain–damaged and control subjects were not affected by changes in the degree of discrepancy. This finding seems to support the hypothesis of a right-hemisphere contribution to the comprehension of emotional prosody, the left hemisphere supposedly being distracted by the semantic content of the sentences.

A second study was conducted in which similar groups of subjects listened to tapes of emotionally intoned sentences that were either filtered to remove the verbal content or nonfiltered. Once again, the right-brain–damaged subjects performed more poorly than the other groups of subjects, whose performance was similar. All three groups were similarly affected by the filtered speech condition, with the comprehension of prosody being worse in sentences that were filtered.

This effect was interpreted by Bowers et al. (1987) as support for the presence of a processing defect in right-brain–damaged patients. When confronting the results of the two studies, Bowers et al. conclude "that right-brain–damaged patients have both a processing defect and a distraction defect that contribute to their poor performance on emotional prosody tasks" (p. 327).

Several comments can be made about this paper. The first concerns the so-called distraction defect hypothesis. Insofar as this hypothesis assumes the existence of an impairment of left-hemisphere language processes, it would have been pertinent to provide some information concerning the linguistic abilities of the left-brain–damaged subjects, in particular, whether subjects were aphasic or not. A second comment has to do with the task requirements. Indeed, the subjects were instructed not to pay attention to the semantic content of the stimuli but rather to concentrate on the intonation. It is hard to imagine that the subjects followed these instructions to the letter. Thus a suitable conclusion to this study would be that the comprehension of emotional prosody, when associated with a verbal message, seems to depend in part on a contribution of the left hemisphere.

The importance of the nature of the task has also been examined in studies on lateral processing of emotional prosody by normal subjects. Safer and Leventhal (1977) evaluated the performance of 36 normal subjects on a task that involved the decoding of emotional information. The stimuli were monaural sentences with verbal emotional content as well as emotional intonation. When the stimulus was presented to the left ear, most subjects based their global appreciation of the emotion on the intonation. However, when the stimulus was presented to the right ear, the emotion was deduced from the content of the stimulus. In a second experiment, subjects were asked to identify the emotional nature of both the verbal content and the intonation. In this case, subjects who heard the stimulus in the right ear performed significantly better. This dual task therefore seemed to favor an involvement of the left hemisphere.

In another study, Ley and Bryden (1982) used dichotic presentations. The stimuli were pairs of semantically neutral sentences, one of which contained an intonation conveying a particular emotion whereas the other contained a neutral intonation. The subjects were required to indicate, in a multiple-choice decision task, both the lexicosemantic content of the stimulus and the emotion conveyed by its intonation. The majority of the subjects (21/31) demonstrated a right ear advantage (left hemisphere) for content and a left ear advantage (right hemisphere) for intonation. Although the authors of these two distinct studies agree as to the involvement of the right hemisphere during the processing of emotional prosodic stimuli, their interpretations differ. Safer and Leventhal (1977) emphasized the nature of the task, whereas Ley and Bryden (1982) emphasized the nature of the stimulus (verbal or intonational emotion) and rejected the effect of any subject strategies: "The laterality effects observed in the present study are related to differential cerebral lateralization rather than to subject strategy effects" (p. 7). Clearly, more research is needed to disentangle the role of each hemisphere in the comprehension of emotional prosody.

Emotional Prosody and Pragmatics

The next two studies emphasize the pragmatic role of intonation. Weniger (1984) as well as Tompkins and Mateer (1985) sought to evaluate the comprehension of emotional prosody in tasks where it changed the meaning of a given context. In everyday conversations, we establish a relationship between intonation and verbal content in order to grasp the intention of the speaker. Thus emotional prosody is used (1) to eliminate the ambiguity of a neutral sentence, by expressing the emotions of the speaker which the literal verbal expression does not convey, (2) to reinforce the meaning of the verbal message (e.g., "You have passed your exam" with a happy tone), or (3) to contradict the literal meaning of the verbal message (e.g., "I am pleased with your work" with an angry tone).

Weniger (1984) evaluated the interpretation of neutral sentences differing in terms of the intonation with which they were spoken. For example, if the sentence "The cage was open" were spoken with a sad tone, this sentence could convey the regret associated with the loss of a familiar animal. However, if the same sentence were spoken with an emotion of fear, this could convey a feeling of possible danger due to the escape of a wild animal. The subjects were instructed to choose the picture associated with the correct interpretation of the intonation (i.e., fear, joy, anger, or sadness). The performance of right-brain–damaged subjects and Wernicke's aphasics was inferior to that of Broca's aphasics and normal subjects. However, Weniger neglected to provide any detailed information concerning these groups of subjects and, moreover, did not mention their abilities to decode the intonation itself or the types of errors that were made. A task such as this seems to require at least three different operations on the part of the subject:[4]

1. Normal decoding of the verbal information. The failure of the Wernicke's aphasics most likely occurred at this level and could be reflected in the percentage of distractors chosen.
2. Adequate decoding of the intonation. Do the subjects really understand whether they are dealing with anger or with sadness? Studies to this effect have suggested that right-brain–damaged subjects make more errors at this level.
3. Determining which of the two sources of information (1) and (2) above is necessary to decide on the appropriate interpretation of the stimulus. If the objective is to analyze this last operation, then it is necessary to establish relationships with the preceding levels. In other words, this study confirms the difficulty of right-brain–damaged subjects to process emotional prosodic information but does not inform as to what extent these difficulties are increased or not by the confrontation of the prosody with a given context.

Tompkins and Mateer (1985) asked their subjects to determine whether an intonation was congruent with a given context. The verbal context itself (i.e., a short paragraph) also contained congruent or contradictory information, for

[4]Picture interpretation itself is likely to cause certain problems.

example, "Nan Smith invited her new neighbour, Mark, to a party. He told hilarious stories, and everyone enjoyed listening. Nan's husband said to her, 'Good decision. He's really the life of the party.'" The experimental group consisted of 18 patients with complex epileptic syndromes (10 patients with a right temporal epileptogenic lesion and 8 patients with a left temporal lesion). Twenty-four college students served as normal controls.

In contrast to the Weniger (1984) study, an additional task was included to evaluate comprehension of the verbal message itself as well as the subjects' capacity to deduce the attitude that it implicated on the part of the speaker. Only congruent paragraphs were understood without any difficulty. In their subsequent analysis of the prosodic task, the authors focused only on these paragraphs.[5] It was found that the right-brain–damaged subjects made more errors than the other two groups of subjects. They also had a tendency to report the existence of a conflict between emotional intonation and content when there was none. Left-brain–damaged and normal subjects performed in much the same way.

A further innovation of this study was the analysis of individual performance within the group of right-brain–damaged subjects, which was found to be heterogeneous. Indeed, the performance of five of the ten right-brain–damaged subjects was similar to that of the normal subjects. More importantly, these were the same subjects who performed within the limits of the normal range on Dodrill's neuropsychological test battery. This battery includes subtests of the Wechsler Adult Intelligence Scale and the Wechsler Memory Test, as well as a subtest of tonal memory from the Seashore scales.

Conversely, the five subjects who failed the prosodic task also obtained below-normal scores on Dodrill's test battery. In addition, tonal memory was poor, whereas the subjects who succeeded in the prosodic task had normal tonal memory. Thus the patients who performed poorly on the neuropsychological tests, and notably on the tonal memory test, were the same who had difficulties extracting prosodic information. These results suggest that the difficulties of this particular group of subjects are not so much the result of the prosodic representation of an emotion or of an attitude as they are of a more fundamental disorder in the processing of a prosodic auditory information (Milner, 1962). The particularity of this population also makes any comparison with previously mentioned studies difficult. However, the methodology of this study exemplifies the problems encountered and the questions raised in previous studies. The initial objective of the authors was to assess the pragmatic role of intonation, but the rigor of their methodology reinstated the problem at a much more elementary level. In this population, there is a strong correlation with disorders that extend beyond the limits of emotional prosody and which concern, more fundamentally, acoustic processing (tonal memory task) and/or a global deficit of intellectual functions. The results of this study, through its use of additional control tasks,

[5]The problems associated with the comprehension of the paragraphs are examined in Chapter 7.

have led to questions concerning other studies in which only prosody has been analyzed, forcibly leading to the conclusion that the disorder is indeed one of prosody. Likewise, the analysis of individual performance reveals the shortcomings of group analysis (see the foregoing discussion of Shapiro & Danly, 1985).

In summary, right-brain–damaged patients have difficulties with the expression and the comprehension of emotional prosody. The disorder of emotional prosodic comprehension seems to begin at the early stage of perceptual decoding of prosodic information and does not seem to depend solely on its emotional nature. Finally, not all right-brain–damaged subjects are affected by these difficulties. It is easy to propose that disorders of global behavior in anosognosic and indifferent patients are also manifest in their expression and their comprehension of emotional prosody (Tucker et al., 1977). However, it is much more difficult to speak of right-brain–damaged subjects as a whole. There are arguments suggesting a certain amount of diversity either in the comprehension (Tompkins & Mateer, 1985) or in the expression of emotional prosody (Shapiro & Danly, 1985). However, we do not yet have the necessary tools to predict the presence of any eventual anatomoclinical correlation. It is also clear that aphasic patients have difficulties with the comprehension of prosody, since the processing of emotional prosody is not performed separately but involves instead the integration of other sources of information, notably verbal.

Linguistic Prosody

An intrinsic component of any verbal message is the linguistic prosody associated with its linguistic structure. Prosody is involved in the structure of the message, such as the intonational contour one uses to stress the interrogative or declarative modality of a statement. The problem then is to determine if this linguistic prosody is dependent on processes controlled by the left hemisphere because of its linguistic function or if it depends on right-hemisphere processes because of its prosodic structure (Zurif, 1974).

Understanding Linguistic Prosody

According to Green and Boller (1974), aphasic subjects should be capable of distinguishing the modality of a statement in spite of having a severe comprehension disorder. For example, they would be able to distinguish an imperative from an interrogative sentence. However, evidence of this type is far too general for one to conclude that linguistic prosody is preserved. Indeed, aphasics seem to rely on cues from the facial expressions or the gestures of the speaker. In order to answer the question raised previously, the study of linguistic prosody requires strict experimental protocols, far from the normal situation in which prosody and the verbal message are intertwined. We begin by reviewing several studies of normal subjects as they reveal two important concepts that will help explain the contradictory findings with right-brain–damaged or aphasic patients: the nature of the task and the type of linguistic prosody studied.

Dichotic Listening in Normal Subjects

Zurif and Sait (1970) as well as Zurif and Mendelsohn (1972) sought to evaluate the role of the intonational contour associated with meaningless, syntactically organized sentences. In one study, Zurif and Mendelsohn replaced the words of a short sentence with meaningless syllables but retained the grammatical morphemes and the function words so as to maintain the syntactic structure (e.g., "Lo ddont jeg ub shavving"). These meaningless sentences were pronounced either with normal intonation adapted to their syntactic structure or with monotonous intonation (i.e., each element was pronounced with the same tone of voice). Subjects were presented with dichotic pairs of similar sentences (i.e., normal or monotonous intonation) and asked to identify one of the two sentences from a list of four choices. A group of 48 normal subjects demonstrated a right ear advantage for sentences with normal intonation but no ear advantage for sentences with monotone intonation. Thus in a linguistic decision task (i.e., identifying a syntactically organized sequence), the right ear advantage is observed only when intonation is normal. In the interpretation of these results, the role of the task is emphasized: "In this view, intonation, although by itself probably processed by acoustic analyzers in the right hemisphere, enhances the left hemisphere's ability to carry out linguistic decision" (Zurif, 1974, p. 394). Stated otherwise, the acoustic parameters of prosody, when in the service of a linguistic task, are processed by the left hemisphere.

Blumstein and Cooper (1974) compared the processing of prosodic intonation associated with linguistic information with that of isolated intonation. In one experiment, the stimuli were dichotically presented sentences expressing one of four modalities: declarative, interrogative, imperative, or conditional (e.g., "It has come. Has it come? Hal come here. If he came"). The sentences were filtered, so that only the intonational contour was intelligible. In one of the tasks, the dichotic sentences were followed by the binaural presentation of an intonational contour. The subjects had to indicate if this contour was identical to either one of the previously presented sentences. The results revealed a significant left ear advantage (right hemisphere). In another task, half of the subjects ($N = 18$) had to select from a series of curves the two which depicted the intonational contours of the stimuli (nonlinguistic task). The other half of the subjects had to identify the sentences corresponding to the intonations (linguistic task). In the nonlinguistic task, there was a superiority of the left ear (right hemisphere). However, a superiority of the left ear was also found in the linguistic task, though it was no longer significant. Thus it seems that there is a right-hemisphere advantage for the discrimination and the identification of filtered intonations except when the response is linguistic (e.g., designating the corresponding sentence), in which case the involvement of the left hemisphere results in no between-ear difference in performance.

In a second experiment, Blumstein and Cooper (1974) sought to determine if the coexistence of verbal information could modify the processing of modality. The same intonations could modify the processing of modality. The same intonations as those of the first experiment were now associated with meaningless filtered or unfiltered syllables (e.g., padaka). The task was to match the dichotically presented stimuli with an intonation contour, which was always filtered. The results of a group of 20 students revealed a left ear advantage (right hemisphere) in the filtered condition and no between-ear difference in the unfiltered condition.

In summary, there is a right-hemisphere superiority for the processing of intonational-based modality when the sentence has lost its verbal component (i.e., filtered conditions) and when the task is not linguistic. When phonological information is present (e.g., syllables) or when the response is linguistic, then the differences between the two hemispheres are no longer significant. Contrary to the Zurif and Mendelsohn (1972) study, Blumstein and Cooper (1974) failed to observe a right ear advantage (left hemisphere) in any of their conditions. Their interpretation of their findings is cautious: "The data from this experiment indicate that, at least, the perception of intonational contours is predominantly a right hemisphere function" (p. 155). Their conclusion supports the view that language perception occurs by way of a parallel and independent processing. The analysis of phonetic and semantic components would seem to be conducted primarily in the left hemisphere, whereas the analysis of prosodic components would appear to take place primarily in the right hemisphere. This notion of independent processing was later confirmed in a dichotic listening study using musical stimuli and digits (Goodglass & Calderon, 1977).

The difference between the results of Zurif and co-workers (Zurif, 1974; Zurif & Mendelsohn, 1972) and those of Blumstein and Cooper (1974) lies in the nature of the verbal stimuli as well as the task. Blumstein and Cooper employed nonsense syllables, whereas the stimuli used by Zurif and Mendelsohn were syntactically organized. In their so-called linguistic recognition task, Blumstein and Cooper instructed their subjects to identify a filtered intonational contour, whereas Zurif and Mendelsohn presented a sentence and its intonational contour as possible choices. In other words, the linguistic "load" seems to be more important in the Zurif and Mendelsohn experiment, which could account for their observation of right ear advantage (left hemisphere). If the methodological differences between these two studies help explain differences in the lateralization of prosodic processing, then it becomes difficult to consider the existence of independent processing by the two hemispheres simply in the terms of verbal versus prosodic. On the basis of these two studies, it seems more reasonable to conclude that the left hemisphere is involved in the processing of prosodic information, but to a varying degree depending on the nature of the relationship between the linguistic material and the intonation. Stated otherwise, the nature of the task and the processing it involves are certainly the most important points to consider in order to understand the involvement of each hemisphere in the processing of linguistic intonation and the differences observed between right- and left-brain–damaged populations.

This issue has been further illustrated in a study of stress placement (Behrens, 1985). The stimuli consisted of noun phrases such as "REDcoat," in which the stress is placed on the first syllable, versus "red COAT," in which the stress is placed on the second syllable. Pairs of such stimuli (e.g., "white cap/redcoat" or red coat/whitecap") were presented dichotically. The subjects were instructed to write down the number ("1" or "2") corresponding either to the stimulus that was stressed on the first syllable or to the stimulus that was stressed on the second syllable. A group of 15 normal subjects demonstrated a right ear (left hemisphere) advantage on this task. In a second experiment, the stimuli were filtered to test stress placement once the phonetic and the semantic components had been eliminated. Eleven of the nineteen subjects who performed the task were not retained for the final analysis since their performance was less than the level of chance set at 60%. As for the eight remaining subjects, there was a significant left ear advantage, thereby suggesting a predominance of the right hemisphere. This change in the ear advantage implies that in the first experiment, the subjects were using strategies based on phonetic or semantic information, thereby explaining the involvement of the dominant left hemisphere. A third experiment fulfilled an intermediate condition between the two preceding experiments. The stimuli from the first two experiments were used to construct a set of bisyllabic stimuli, which provided phonetic information but which were also meaningless. The results of 15 subjects revealed a nonsignificant left ear advantage.

From these three experiments, Behrens (1985) was able to show a cumulative effect in the lateral processing of stress. If phonetic and semantic information is present, then performance is indicative of preferential processing by the left hemisphere. If linguistic support is present but meaningless, there is no significant difference between the two hemispheres. Finally, if only intonational cues are present, then processing involves the right hemisphere. The results of this study suggest that in everyday life, normal language is perceived as a whole (i.e., intonation is perceived at the same time as phonetics and semantics) and is treated as such, preferably by the left hemisphere. The left-hemisphere processes become involved in prosodic processing whenever the prosodic features convey linguistic meaning, whereas preferential processing by the right hemisphere occurs only when the intonation is separated from its linguistic component.

The capacity of the left hemisphere to process prosodic information associated with linguistic content is also supported by the findings of Hartje, Willmes, and Weniger (1985). Thirty normal subjects were instructed to solve a verbal prosodic task. The stimuli were dichotic pairs of syllabic triplets (e.g., gamala–dabala) presented with a particular intonation. The subjects were given four possible solutions: two correct responses (i.e., each triplet and its associated intonation) and two erroneous responses containing the target triplets but with inversed intonation curves. With respect to the stimuli presented to the left ear (right hemisphere), the subjects made more errors of the dominant type, that is, they recognized the intonation but they associated it with the verbal information presented to the right ear (left hemisphere). This finding is therefore compatible with the limited verbal capacity of the right hemisphere. The intonation is processed by the right hemisphere and is combined with the verbal information

:essed by the left hemisphere. However, the inverse pattern of errors was not served for the stimuli presented to the right ear. Stated otherwise, the left hemi-ɔhere seems capable of processing both verbal information and intonation. Finally, it should be noted that although modality and emotional intonations were used in this study, the authors did not differentiate these two types of stimuli.

In summary, the evidence from dichotic listening studies of the identification of linguistic stimuli associated with particular intonations (Behrens, 1985; Hartje et al., 1985) favors a joint processing of such information by the left hemisphere, with no tendency to transfer the processing of intonation itself to the right hemisphere. It should also be noted that the level of complexity of the tasks used in these two studies, which is not that of most of the studies with brain-damaged subjects, could in itself result in a contribution of the left hemisphere.

Studies of Brain-Damaged Subjects

The methodology of the studies just described was also used in a study with brain-damaged subjects (Hartje et al., 1985). However, this particular study has two major limitations. The first concerns the use of dichotic listening tasks with subjects presenting a unilateral lesion (Bradshaw et al., 1986). The second concerns the fact that the authors of the study confound linguistic prosody and emotional prosody. In spite of these limitations the results of the study indicate that in a complex task, in which both prosodic and verbal processing are involved, the left hemisphere can ensure efficient prosodic processing when the right hemisphere is damaged. The stimuli were dichotic pairs of syllabic triplets presented with a particular intonation. Following each presentation, the subjects were given a list of six triplets and had to identify the two stimuli. They were also given six contours (or scenes) and asked to select the two contours depicting the intonations of each stimulus pair. The subjects were 15 aphasics, 9 right-brain–damaged patients, and 10 normals. As expected, the aphasics identified intonations with their right hemisphere, whereas right-brain–damaged subjects identified syllables with their left hemisphere. Furthermore, the aphasic subjects identified a greater number of intonation contours with their right hemisphere (left ear) than did the right-brain–damaged subjects with their left hemisphere (right ear). An even more important finding concerning the identification of prosody was that the performance of the left hemisphere of the right-brain– damaged subjects did not differ significantly from that of normals. In other words, following a lesion of the right hemisphere, the left hemisphere seems capable of identifying intonational contours in a dual task (i.e., syllabic and intonational identification) in which intonation is processed at the same time as verbal information.

The following studies have been grouped according to the type of linguistic prosody involved (i.e., lexical stress, emphatic stress, and sentence intonation).

Understanding Lexical Stress

In English, the placement of stress on either the first or the second syllable allows one to distinguish between "IMport" and "imPORT" or "REDcoat" and "red

COAT." Comprehension of lexical stress was first evaluated in left-brain–damaged patients. These studies provided information as to whether the intact right hemisphere was sufficient to allow for the normal understanding of lexical stress. Blumstein and Goodglass (1972) had their subjects listen to a series of words, such as redcoat, and asked them to select from a series of four pictures the one corresponding to the word. In the example, the correct picture would be that of a British soldier of the eighteenth century, the other pictures consisting of a red coat and two distractors (i.e., a red cap and a porter dressed in red).

In this study, the lexical stess errors of aphasics ($N = 17$) were found to be similar to those of normals ($N = 13$). However, this finding does not warrant, in our opinion, the following conclusion: "It has also been shown that the comprehension of stress contrasts . . . is remarkably well preserved in aphasia" (Blumstein & Cooper, 1974, p. 156). In reality, the results show that Broca's and Wernicke's aphasics, as well as normals, made numerous errors. The task seems to have produced a floor effect, thereby excluding the conclusion that aphasics are capable of decoding lexical stress. A similar study in German (Weniger, 1978) revealed that errors due to poor comprehension of the placement of lexical stress were more frequent in aphasics than in normals.[6] It should be noted, however, that the task was more difficult than the one used by Blumstein and Goodglass (1972), since twice as many pictures were used (eight pictures).

In a somewhat similar study, Baum, Daniloff, Daniloff, and Lewis (1982) examined the comprehension of lexical stress in sentences such as "She is home sick" and "She is homesick." A group of Broca's aphasics ($N = 8$) made significantly more errors than a group of normal subjects, and even more so as their aphasia increased in severity. It should also be added that the Broca's aphasics had difficulty with the contrasting stress, which signals the boundary between two morphemes (e.g., "It's a gray day" versus "It's a grade A").

Emmorey (1984) used the same procedure as that of Blumstein and Goodglass (1972) but also included right-brain–damaged patients. She compared the performance of aphasic ($N = 15$), right-brain–damaged ($N = 7$), and normal ($N = 22$) subjects. The success rate of the right-brain–damaged patients (76%) was similar to that of the normal subjects (87%) and significantly better than that of the aphasics, whether they were fluent (55%) or nonfluent (62%). In another study, Weintraub, Mesulam, and Kramer (1981) found that right-brain–damaged subjects made significantly more stress-placement errors than normal subjects. It should be noted that with a similar number of stimuli, the success of the right-brain–damaged patients was almost identical (75%) to the results reported by Emmorey (1984).

The results of these studies seem to suggest, notwithstanding the limitations concerning the Blumstein and Goodglass (1972) study, that left-brain–damaged aphasics do experience difficulties in understanding the acoustic cues involved in lexical stress. Therefore, the integrity of the right hemisphere is not sufficient to ensure normal performance on these tasks. The comparison of right- and left-

[6]For example, "EINgriff" means "operation," whereas "ein GRIFF" means "a sleeve."

brain–damaged subjects (Emmorey, 1984) suggests, in fact, a necessary contribution of the left hemisphere.

Emphatic Stress and Intonation-Based Modality

Weintraub et al. (1981) focused their attention on emphatic stress and intonation-based modality in a discrimination task. The subjects were asked to determine whether two sentences were identical. The sentences differed (1) by the emphatic stress, which was either on the first word or on the last word (e.g., "*Steve* drives the car" versus "Steve drives the *car*"), or (2) by the intonation contour associated with a given modality, which was either declarative or interrogative (e.g., "Margo plays the piano" versus "Margo plays the piano?"). It was found that right-brain–damaged subjects ($N = 9$) made more errors than normal subjects ($N = 10$). The impairment of the right-brain–damaged subjects appears to concern the actual perceptual decoding of prosody, independently of its linguistic function. This hypothesis is all the more plausible since these subjects also had difficulties discriminating emotional prosody, as discussed previously. Another argument to support this hypothesis would be the observation of a positive correlation between scores for emotional or linguistic prosody and a deficit on an auditory discrimination test. However, no such evidence has yet been provided. The results of the Weintraub et al. (1981) study suggest that several different types of linguistic prosody (e.g., lexical stress, emphatic stress, and accentuation of the intonation contour) were at the source of the difficulties of the right-brain–damaged subjects. However, it is not known if these difficulties were of the same nature or if they were correlated to one another.

Heilman et al. (1984) examined the performance of 8 right-brain–damaged subjects (5 of whom had hemineglect), 9 aphasics (5 nonfluents and 4 fluents), and 15 control subjects on an identification task. The stimuli were filtered sentences from each of the following modalities: declarative, imperative, or interrogative. The task was to identify the punctuation mark associated with each of the sentences (".," "!," or "?"). The success rate of the right-brain–damaged subjects (60.3%) was similar to that of the aphasics (59.8%) but significantly lower than that of the normals (88.8%).

In other words, when the intonation contour is not associated with a linguistic structure, it seems to create quantitatively similar difficulties for right-brain–damaged and aphasic subjects. However, it could be assumed that two distinct mechanisms are involved. The failure of the right-brain–damaged subjects is compatible with the perceptual hypothesis mentioned earlier. As for the failure of the aphasic subjects, this could be attributed, in part, to the nature of the task, which requires that different intonation contours be confronted with the corresponding modality punctuation marks.

The results of these studies can be summarized succinctly. Linguistic prosody is disturbed whatever the lateralization of the lesion (1) in cases of a left-hemisphere lesion because it involves linguistic processing and (2) in cases of a right-hemisphere lesion because it involves prosodic processing. These findings are compatible with the evidence from studies of normal subjects (see Behrens,

1985; Blumstein & Cooper, 1974; Zurif, 1974; Zurif et al., 1970, 1972). Thus the processing of linguistic prosody should not be conceived as depending on an exclusive set of processes enclosed within a *single* "module." Moreover, the results of these studies suggest that the contribution of the right hemisphere depends on (1) the type of linguistic prosody considered, (2) the nature of the relationship between linguistic prosody and verbal segments, and (3) the nature of the task required to test different types of linguistic prosody.

From this viewpoint, the model proposed by Grant and Dingwall (1985) adds interesting perspectives, at least with respect to the English language. First, it postulates that lexical stress if linked to a given linguistic segment, whereas intonation-based modality is more independent and concerns the sentence as a whole. The hypothesis advanced by these authors is that the left hemisphere is all the more involved as prosody is related to a given segment (stress), whereas the right hemisphere is all the more involved as prosody occurs over a longer period of time (and therefore involves a greater number of segments). Grant and Dingwall used a discrimination task to evaluate the comprehension of shifts in grammatical class as a function of the placement of lexical stress. Thus the word "import" in English could be either a noun or a verb, depending on whether the stress is placed on the first or second syllable. Right-brain–damaged subjects ($N = 9$) performed in much the same way as aphasics ($N = 9$). However, the performance of the right-brain–damaged subjects was significantly less than that of normals. In a second task, the subjects were required to discriminate between the intonation-based modality of two sentences. Right-brain–damaged subjects made significantly more errors than aphasics, whether or not the sentences were filtered, and the performance of both of these groups was significantly worse than that of normal subjects.

In summary, right-brain–damaged patients obtained lower scores than normals for both types of linguistic prosody and obtained lower scores than aphasics only for prosody marking intonation-based modality. The authors concluded that each hemisphere is involved to a varying degree, depending on the type of linguistic prosody. The findings are compatible with this hypothesis. Indeed, the right hemisphere is more involved in the decoding of intonational contour than it is in the decoding of lexical stress. The Grant and Dingwall (1985) model requires more study, but it should be recognized that it is compatible with the results of previously mentioned studies. A parallel should be drawn between this viewpoint and current knowledge concerning the lateral processing of consonants and vowels. Several studies have suggested that rapid and local processing of acoustic information is carried out by the left hemisphere, whereas processing over a longer period of time is carried out by the right hemisphere (e.g., Perecman, 1983; Perecman & Kellar, 1981; Yeni-Komshian & Rao, 1980). Moreover, the Grant and Dingwall model is similar to the distinction that has been made between local and global processing of melodic stimuli (Peretz, Morais, & Bertelson, 1987; also alluded to by Van Lanckner, 1980).

To summarize, studies concerning the understanding of linguistic prosody have shown that (1) there is sufficient evidence to reject the idea of a parallel and independent representation of linguistic prosody in terms of a dichotomy such as the verbal–nonverbal; (2) part of the difficulty of right-brain–damaged subjects can be attributed to a disorder in the perceptual decoding of prosody itself; and (3) the nature of the processes involved within each hemisphere are probably dependent on the type of prosody in a given verbal message (e.g., stress accentuation or intonation).

Production of Linguistic Prosody

The main objective of the next group of studies reviewed here has been to describe the acoustic parameters involved in the production of prosody. Due to methodological constraints, studies of the production of prosody do not allow for an analysis of spontaneous (e.g., narrative or conversational) speech. Therefore, prosodic production has been evaluated by way of reading tasks, repetition tasks, or answers to specific questions. Some of the studies have provided a global analysis of prosodic production (e.g., Dordain et al., 1971), whereas others have focused on more specific aspects (e.g., Emmorey, 1984).

In 1971 Dordain et al. conducted a clinical and acoustical analysis of the prosodic productions of 17 right-brain–damaged subjects presenting diverse etiologies (six glioblastomas, one meningioma, seven ischemias, and three hematomas). The clinical analysis of their productions revealed difficulties in the control of voice intensity, which was judged as being too loud in 15 of the 17 subjects. The subjects were unable to increase or reduce their voice intensity on demand (e.g., nine patients were unable to murmur). An acoustic analysis of the frequency parameter was performed, using the glottographic waves produced during a reading or a repetition task. The following parameters were calculated: (1) the mean fundamental frequency (F_0), (2) the variation of this frequency, and (3) a coefficient of variation based on the relationship between the variance of F_0 and its mean value. The acoustic measures obtained for the right-brain–damaged subjects were compared with those of 42 normal subjects. Only 3 of the 17 right-brain–damaged subjects demonstrated a variation of frequency similar to that of normal subjects. As for the other right-brain–damaged subjects, this variation was either greater ($N = 5$) or less ($N = 9$) than that of normals. The results of this study are also interesting from a methodological point of view. For example, the authors found no correspondence between their clinical impressions and the results of the acoustic analysis. In other words, a voice that was judged to be normal could turn out to be quite variable in terms of its frequency. Conversely, a voice that was judged to be monotonous could have near-normal acoustic parameters. In summary, the two main findings of this study are (1) the great variability of certain acoustic parameters in the prosodic productions of unselected right-brain–damaged subjects and (2) the absence of any correlation between clinical impressions and objective acoustic measures.

Weintraub et al. (1981) examined the repetition of sentences, which differed either in terms of the intonational contour of modality (i.e., declarative or interrogative) or in terms of the placement emphatic stress (i.e., on the subject or on the object) (e.g., "*Steve* drives the car" versus "Steve drives the *car*"). The degree of similarity between the intonation of the examiner and that of the subject was evaluated by a judge. The results revealed that a group of right-brain–damaged subjects (N = 9) made more errors than a group of normal subjects (N = 10). However, the results for each type of linguistic prosody were not provided. In reality, it is difficult to conclude that a production order was involved, since the same subjects also demonstrated difficulties with the comprehension and the discrimination of intonation, as noted earlier. In a second task, Weintraub et al. evaluated the production of emphatic stress: two sentences were read aloud to the subjects, who were then asked a question that called for the placement of emphatic stress either on the subject or on the object.[7] A judge listened to the subjects' responses to determine whether the placement of emphatic stress was correct. Once again, right-brain–damaged subjects made significantly more errors than normals.

Cooper, Soares, Nicol, Michelow, and Goloskie (1984) measured the fundamental frequency as well as the time taken to read a series of sentences which differed in length and structure. One block of stimuli consisted of 12 single-clause sentences of variable length, depending on either the number of syllables or the number of words. A second block of stimuli consisted of 18 two-clause sentences, also of variable length. The fundamental frequenty (F_0) was measured for the first and last words of each clause. With respect to the sentences with a single clause, right-brain–damaged subjects (N = 4) and aphasics (N = 5) had higher F_0 values than normals. However, all of the subjects demonstrated higher F_0 values at the end than at the beginning of each clause. As for the sentences containing two clauses, the tendency of the right-brain–damaged subjects to have higher F_0 values was confirmed. Individual analyses revealed greater variability in left- than in right-brain–damaged subjects with respect to the mean fundamental frequency. As for the mean reading times (i.e., for the different groups and the different sentence lengths), these were found to be somewhat similar in the normal and right-brain–damaged subjects but were significantly longer in the aphasic subjects. The results of this study suggest, therefore, that disorders of the fundamental frequency and duration are greater in left-brain–damaged than in right-brain–damaged subjects.

To evaluate the capacity of brain-damaged subjects to produce lexical stress contrasts, Emmorey (1984, 1987) asked subjects to read aloud component noun phrases such as "redcoat" and "red coat" while the corresponding picture was presented. Fourteen judges were asked to listen to these productions and decide if they denoted the compound word (redcoat) or the noun phrase (red coat). The acoustic measures (i.e., F_0 and duration of productions) were correlated with the evaluations of the judges. All but one of the right-brain–damaged subjects (N = 7), and all but two[8] of the normal subjects (N = 22), were capable of modifying both the pitch and the duration to produce a difference perceptible by the judges.

[7]For example, "The man walked to the grocery store. The woman rode to the shoe store." If asked "Who walked to the grocery store, the man or the woman?," the correct answer would be "The *man* walked to the grocery store," with placement of emphatic stress on "man."

[8]All but one in the Emmorey (1987) paper.

In contrast, aphasics ($N = 15$) had difficulties changing these acoustic parameters. On the basis of these findings, Emmorey (1984) concluded that the acoustic parameters of lexical stress are controlled by the left hemisphere. This conclusion is compatible with the fact that lexical stress is the prosodic element most closely linked to the lexical representation of words.

Behrens (1986) compared the production of lexical stress and emphatic stress in eight right-brain–damaged and seven normal subjects. The production of lexical stress was tested with stimuli such as "redcoat" and "red coat." The production of these stimuli was elicited by the presentation of a picture associated with a short story. Productions of emphatic stress were obtained by way of a technique similar to the one used by Weintraub et al. (1981): the subject's response to specific questions required the placement of emphatic stress either on the subject or on the object. Three judges listened to each of the productions and determined, in the first task, the nature of the segment produced (i.e., compound word or noun phrase) and, in the second task, the placement of emphatic stress. Acoustic measures were taken for each syllable (i.e., amplitude, fundamental frequency, and duration of the intersyllabic pause). First, the results were analyzed to determine if the acoustic measures were correlated with the evaluations of the judges; that is, is there a relationship between the acoustic parameters involved and a judge's evaluation of the production of lexical or emphatic stress? For example, does the fundamental frequency of the first syllable change significantly as a function of whether it is stressed? The analysis of individual results revealed that the correlating acoustic parameters varied from one subject to another (e.g., for one subject it was the amplitude, for another the pause, whereas for a third both parameters were correlated). The two normal subjects in whom no acoustic measure was correlated with stress production were the same who obtained the lowest scores from the judges. A similar analysis of the production of the eight right-brain–damaged subjects revealed no correlations in five of them.

The correlations between acoustic data and emphatic stress were higher than the correlations for lexical stress. Each normal subject had at least one acoustic cue correlated with stress, as did six of the right-brain–damaged subjects. The difference between the two types of prosodic stress within the group of right-brain–damaged subjects was also reflected in the judges' scores: 77% success for emphatic stress versus 66% for lexical stress. An interesting finding was that the two patients who showed no correlation for emphatic stress did show a correlation for lexical stress. All told, each right-brain–damaged subject demonstrated a certain capacity to mark stress in at least one of the two tasks. The fact that performance for the two types of stress varied in much the same way in both groups of subjects suggests that right-brain–damaged subjects are just as sensitive as normals to the greater demand of acoustic contrast in emphatic stress versus lexical stress.

In summary, the results of this study suggest the following: (1) there is no simple, homogeneous acoustic behavior to mark stress, (2) this diversity is also observed in right-brain–damaged subjects, and (3) the contribution of the right hemisphere is probably not the same for the different types of linguistic prosody (i.e., lexical versus emphatic stress).

Shapiro and Danly (1985) examined the production of linguistic prosody in three subgroups of right-brain–damaged subjects. These subgroups were formed on the basis of anatomical criteria: the anterior group ($N = 3$, frontal lesion), central group ($N = 3$, prerolandic and parietal lesion), and posterior group ($N = 5$, temporoparietal lesion). The prosodic productions of these subjects were compared with those of five aphasics and five normal subjects.

Acoustic parameters were measured while the subjects read a target sentence. An introductory paragraph, describing a particular context, prompted the subjects to read the target sentence either as a declarative sentence or as an interrogative sentence, and with either a happy or a sad intonation. The measures of mean F_0 values at peaks and valleys (and their sum) revealed no significant differences among right-brain–damaged subjects, aphasics, and normals. Comparisons with the three right-brain–damaged subgroups revealed a tendency of the anterior and central subgroups to present lower values than normals and aphasics, whereas the posterior subgroups had higher values. Furthermore, the anterior and central subgroups evidenced less variation in F_0 than the posterior subgroup, aphasics, and normal subjects. Finally, the difference between the extreme (F_0) values was smaller for both the anterior and the central subgroups than for the other groups of subjects.

The results of the acoustic analyses for the anterior and central subgroups (i.e., less tonal variation, smaller range between extreme values of F_0) are coherent with the flat, monotonous speech usually observed in right-brain–damaged subjects. Conversely, the posterior subgroup had a tendency to use higher frequencies, with greater variation and greater range. Despite this apparent dissociation, Shapiro and Danly (1985) interpret their findings cautiously: "While the present data suggest a dissociation between the anterior and posterior portions of the right hemisphere in the control of speech prosody in general, this distinction may be overly simplified" (p. 32). Indeed, the difference between the three subgroups is not based on criteria attributable to any particular model of prosody. No criteria can explain the similarities in the results of the anterior and central subgroups nor their differences with respect to the posterior subgroup. If this anatomically based distinction does indeed correspond to a difference in the functional organization of prosodic expression, one would expect to observe homogeneous behavior among the subjects of a given subgroup. In reality, the results revealed a large amount of intragroup variability, a finding that is compatible with previously mentioned studies (e.g., Behrens, 1986; Dordain et al., 1971).

Finally, Shapiro and Danly (1985) observed similar acoustic characteristics for both linguistic prosody and emotional prosody. This would appear to suggest the presence of a primitive disorder of prosodic expression subsequent to a lesion of the right hemisphere: "These findings . . . suggest that damage to the right hemisphere alone may result in a primary disturbance of speech prosody that is not necessarily linked to the disturbances in affect often noted in right-brain–damaged populations" (p. 33). This hypothesis is questionable for several reasons (Ryalls, 1986): (1) the small number of patients involved in the study limits its conclusions, (2) the variability in the results makes it difficult to attribute all of the observed differences to a hypothetical primary disturbance, and (3) the acoustic changes noted are not in any way specific to a right-hemisphere lesion since they have also been observed in aphasia (e.g., Cooper et al., 1984; Emmorey, 1984). It seems more appropriate, given the current state of knowledge, and considering the acoustic parameters used, to acknowledge the existence of a certain heterogeneity within the right-brain–damaged population, which does not seem to be related to any anatomical division, at least not at this stage.

In this respect, a study conducted by Ryalls, Joanette, and Feldman (1987) with a much larger number of right-brain–damaged patients ($N = 19$) produced some more convincing evidence. The authors used a repetition task and analyzed the following parameters: mean fundamental frequency, mean deviation of fundamental frequency, mean frequency range, frequency curve across the whole sentence, range between the extreme frequencies, and duration. None of the measures revealed any significant differences between right-brain–damaged ($N = 19$) and normal subjects ($N = 9$). In addition, no significant differences were found within the group of right-brain–damaged subjects as a function of whether the lesion site was pre- ($N = 8$) or retrorolandic ($N = 11$). There was only a tendency for early postonset subjects to have more monotonous productions, thus raising the question of a resolving laryngeal dysarthric component. This study also raises another issue concerning the type of task used to analyze the production of linguistic prosody. Indeed, a repetition task was used, whereas previous studies were based on a reading task (e.g., Behrens, 1986; Shapiro & Danly, 1985).

In summary, one should avoid making any extrapolations with respect to the production of linguistic prosody. It is reasonable to assume that a right-hemisphere lesion can result in a disturbance of this type of prosody. It is also reasonable to assume that this disturbance has a differential effect on the various types of linguistic prosody (Behrens, 1986). However, the diversity of some of the findings has yet to be explained (Behrens, 1986; Dordain et al., 1971; Shapiro & Danly, 1985). Finally, there is nothing in the various studies reviewed here that would authorize the identification of a group of acoustic parameters to help differentiate the disturbances observed in certain right-brain–damaged subjects from those observed in aphasic subjects (Cooper et al., 1984; Emmorey, 1987).

Conclusion

On the basis of the studies reviewed in this chapter, we obviously know more about the right hemisphere's role in prosodic comprehension than we do about its role in prosodic production. With respect to prosodic comprehension, the evidence again suggests that disorders resulting from lesions of the right hemisphere are attributable, in part, to a perceptual disorder of prosodic decoding, independently of its linguistic or emotional function. The involvement of the right hemisphere in the comprehension of linguistic prosody is well established. However, its importance is dependent on the type of linguistic prosody involved (i.e., lexical or emphatic stress). The distinction raised by Grant and Dingwall (1985) between local and rapid processing versus global and shorter processing offers an interesting perspective for subsequent studies of the respective role of each of the two hemispheres in the comprehension of linguistic prosody.

7 Pragmatics

Pragmatics is a field of study only recently established at the crossroads of several other disciplines—philosophy, psychology, and sociology—each of which has provided a particular definition of what pragmatics is all about. For the present discussion, we define pragmatics as the study of relations between language and the contexts in which it is used (Davis & Wilcox, 1985; Foldi, Cicone, & Gardner, 1983). "Context" is used here in reference both to the subject (e.g., his emotional state, his status with respect to his interlocutor) and to the particular communication situation in which he is engaged (e.g., conversation, reading).

The dissociation between linguistic skills and the actual use of language, or communication, in context is exemplified most dramatically in aphasia. Indeed, when the relatives of an aphasic patient declare "He understands everything," when in fact the aphasiological assessment reveals important linguistic deficits, they are referring to the pragmatic dimension of verbal communication. The richness of the context of a conversation has nothing in common with the tasks usually used in aphasiological assessment (e.g., picture pointing, semantic word matching). When engaged in an actual conversation, the aphasic patient uses gestures, facial expressions, or prosody to convey a message, and this with all the more efficiency as the amount of knowledge shared with his interlocutor is important and does not require explicit information.

In one of the many studies that have attempted to verify this aptitude to communicate in spite of severe aphasia, it has been shown that when the context forces them to do so, aphasics are capable of using gestures, facial expressions, or prosody to request what they need (Prinz, 1980). The results of another study suggest that in spite of a severe comprehension disorder, aphasics are able to recognize the intention of the examiner (Green & Boller, 1974). Whereas the content of aphasics' verbal responses to commands or to questions was often erroneous, the form of the response often seemed appropriate. For example, aphasics would respond to a command (e.g., "Get up!") by executing an action, whereas they would attempt to answer verbally a question requiring a verbal response (e.g., "Do you wear a tie?"). Thus in spite of their numerous difficulties in mastering the elementary structures of language (e.g., phonology, syntax, and semantics), aphasics are able to make use of certain contextual elements for the purpose of communication.[1]

Conversely, despite relatively well-preserved linguistic abilities, the communication behavior of right-brain–damaged subjects in certain respects attests to an inadequacy between their language and the context in which it is used. Their remarks can be inappropriate to the topic of the conversation and have no immediate link with the questions asked. Their speech is filled with circumlocutions and punctuated by frequent digressions. Facial expressions and prosody seem to be unexpressive or unadapted to the emotional context. They also tend to use puns that are unfamiliar to their audience or entirely out of context (Critchley, 1962; Gainotti, 1972; Weinstein, 1964; Weinstein & Kahn, 1955; Weinstein & Keller, 1963).

This summary of the semiology now typically associated with right-brain damage is sufficient to warrant a description, in pragmatic terms, of communication disorders in right-brain–damaged patients. In addition to this general description, it was emphasized in Chapter 2 that early studies recognized the problems right-brain–damaged subjects had in any verbal task that required them to group words or sentences into a coherent context. These tasks included solving anagrams (Cavalli et al., 1981; Marcie et al., 1965), completing sentences (Eisenson, 1962, 1973; Rivers & Love, 1980), and interpreting sentences (Hier & Kaplan, 1980).

Converging arguments from the level of general communication behavior, as well as from the results of more specific verbal tasks, have justified the interest given to the in-depth study of the relationships between language and context in right-brain–damaged subjects. Furthermore, right-brain–damaged subjects offer the possibility of examining the use of context in the absence of gross elementary impairments of language. Indeed, the absence of any major disorder of linguistic functioning allows the use of complex tasks, such as storytelling and understanding humorous stories. With respect to this last aspect, attention must be drawn to a characteristic common to all of the studies reviewed in this chapter: they are based on tasks involving the presentation of linguistic material (occasionally iconographic) and generally are not concerned with the spontaneous communication of right-brain–damaged patients. Although this might appear somewhat paradoxical, the analysis of spontaneous behavior, which is ideal from a theoretical point of view, faces two methodological problems that are difficult to avoid: (1) to make intersubject comparisons, it is necessary to have a common reference,

---

[1]Apart from these two studies of a rather general nature, the objective of the various experimental and therapeutic applications of pragmatics in aphasiology has been to evaluate the level of competence for the different means of communications (e.g., gestures, facial expressions, and prosody) and the effective use of these different means as a function of the context (Davis & Wilcox, 1985; Feyereisen & Seron, 1982a, 1982b; Foldi et al., 1983; Morin, Joanette, & Nespoulous, 1986).

and (2) spontaneous speech, such as conversation, is based on a number of mechanisms inextricably intertwined. To analyze these mechanisms with any precision, it is necessary to use tasks that are structured in a very specific manner. For example, the analysis of discourse abilities is usually based not on a conversation but on a story elaborated from a series of pictures. In this way it is possible to study quantitatively and qualitatively both the form and the content of discourse (Joanette, Goulet, Ska, & Nespoulous, 1986). Similarly, to study in detail how subjects interpret this information, it is not enough to ask them questions about the content of a story; specific tasks must be used to evaluate the various stages of this interpretation (Brownell et al., 1986).

Before examining these tasks, a simple sentence such as "The window is still open" can be used to illustrate the principal notions underlying language- context relationships, which will be clarified further with each of the topics examined. By now, the reader has already interpreted the example given above. Any normal listener (or reader) could interpret this message. However, this interpretation might differ depending on the contexts that it conjures up. Depending on these contexts, the literal expression "The window was open" can take on several different meanings. For example, (1) scolding the person who should have closed it, (2) commenting on the temperature of a room, (3) suggesting through a metaphoric message that some negotiations are still possible, or (4) commenting sarcastically on a disappointing tourist attraction such as a castle in ruins. By way of this example, we wish to underline the fact that communication is not limited to a linguistic analysis of juxtaposed words adhering to an adequate syntacticosemantic organization. Understanding what another person says during a conversation, or understanding the plot of a story, means being able to go beyond the simple literal decoding of words in order to deduce the intention of the speaker or the motivation of the story's characters. To follow a conversation, to understand a text or a joke, requires the establishment of relationships between several elements and their interpretation with respect to a given context.

To ensure that a conversation does not become a dialogue of the deaf, or that a text remains a text and not a simple sequence of sentences, not only must the phonological, syntactic, or semantic aspects of language be mastered, but the rules governing communication must also be respected. A conversation or a text must respect certain rules if a coherent interpretation is to be possible. One of the aims of pragmatics is to formalize the rules which help determine, for example, why a story is perceived as coherent (Charolles, 1976, 1978), why a statement is perceived as humorous (Suls, 1983), why the intention of the speaker is accessible (Searle, 1969), or why the metaphoric meaning of an expression is understood (Gibbs, 1985; Searle, 1982). The use of these forms of reference in neuropsychology is relatively recent and many of the components of communication have yet to be studied in right-brain–damaged patients.[2] In beginning this chapter

[2]Prosody is the component of verbal communication that has been studied the most in right-brain–damaged subjects. The importance given to this question in the literature and the specificity of the conceptual and methodological problems it poses were important in our decision to devote an entire chapter to it (see Chapter 6).

with a review of discourse abilities, we will see to what extent right-brain–damaged patients are able to respect and/or use the usual rules in the elaboration of narrative speech. Next we will focus our attention on the interpretation of discourse, followed by the interpretation of humorous speech. Finally, we will review a number of studies of indirect language acts, on the one hand, and of metaphors, on the other hand. These studies will enable us to examine the difficulties encountered by right-brain–damaged patients in grasping the nonliteral meaning of information.

Discourse Abilities

When reading the various descriptions of verbal communication impairment in right-brain–damaged patients, it becomes apparent that the authors' conclusions are based on very different levels of description. For example, researchers such as Goldstein (1948), Gainotti (1972), and Ross and Mesulam (1979) emphasize the apparent emotional indifference of these patients. Others such as Eisenson (1973) and Weinstein (1964) emphasize the lack of organization of discourse which appears desultory, as well as its sometimes bizarre informative content.

The first objective of a systematic analysis of discourse[3] would be to go beyond the clinical impressions in order to describe, in as much detail as possible, the various aspects of discourse abilities that may be affected in right-brain–damaged patients. Since the mid-1970s numerous psycholinguistic models have offered descriptions of the structures and the processes underlying normal discourse. These so-called text grammars (Charolles, 1978; Kintsch & Van Dijk, 1975; Van Dijk & Kintsch, 1983) could provide reference models for a detailed description of discourse impairment in right-brain–damaged right-handers. One of the first objectives of such an undertaking would be the identification of a semiology of discourse impairment following a lesion of the right hemisphere. But beyond such a description, the ultimate objective would be to identify those basic cognitive processes—may they be a part of memory, language, or any other cognitive function—whose impairment would result in discourse impairment.

We begin by examining one of the first hypotheses to stem directly from the clinical observation that the speech of right-brain–damaged subjects reflects a fundamental disorder of the expressive, emotional function of communication. Clinicians have long recognized the contrast between the indifference, or euphoria, of right-brain–damaged patients concerning their illness and the nonindifferent catastrophic reaction of Broca's aphasics (Gainotti, 1972; Goldstein, 1948). This classical observation is at the origin of a number of studies in which the expressive and comprehension performance of right-brain–damaged patients

[3]The term discourse refers to a group of sentences such as in a conversation or a story. However, since it is generally admitted in psycholinguistics that the structural properties of a text and those of a conversation are essentially the same, the comprehension of written texts is often used to make conclusions concerning the capacities to understand conversational speech (Davis & Wilcox, 1985; Van Dijk & Kintsch, 1983).

has been compared with that of aphasics for the emotional aspects of gestures, facial expressions, or prosody (Feyereisen & Seron, 1982a, 1982b). In light of the inadequacy of the emotional behavior of right-brain–damaged subjects, in both their verbal and their nonverbal behavior, it seems legitimate to hypothesize that certain aspects of discourse impairment are but a consequence of a more generalized emotional or affective disorder.

This hypothesis was examined in an experiment in which the recall of an emotionally laden story, the content of which dealt with "material" related to illness, was compared with that of a story containing an emotionally neutral content (Wechsler, 1973). Recall of the emotionally laden story produced significantly more digressions or comments in right-brain–damaged subjects ($N = 17$) than in aphasic ($N = 10$) or nonaphasic ($N = 7$) left-brain–damaged subjects. Conversely, recall of the story with an emotionally neutral content revealed no difference between the two groups of subjects. The results of this story recall task therefore confirm the effect of emotional content related to illness, which had previously been suggested in the naming performance of right-brain–damaged patients (Weinstein & Keller, 1963).

These two studies highlight the fact that an emotional context can modify the anticipated responses (e.g., naming, Weinstein & Keller, 1963) or can result in discourse recall performance that is not as expected (Wechsler, 1973). This observation tends to suggest that beyond the clinical impression of the emotional inadequacy shown by these patients, a certain access to the emotional value of information still persists and seems to modulate their behavior. However, the methodology used in these studies does not allow any interpretation whatsoever of how right-brain–damaged subjects process the emotional content of discourse and even less of the role of a possible emotional impairment in the discourse abilities of right-brain–damaged subjects.

The processing of emotional information was examined further in a study conducted by Cicone, Wapner, and Gardner (1980). The subjects were asked to match either a picture or a sentence with one expressing a similar emotional state. For example, the picture of an individual threatened with a revolver had to be matched with that of an individual sinking in quicksand since both situations supposedly conveyed an emotion of fear. As expected, left-brain–damaged subjects ($N = 19$) performed more poorly on the verbal portion (sentences) than on the nonverbal portion (iconographic) of the task. A more surprising result, though, was the impaired performance of the right-brain–damaged subjects ($N = 21$) on the sentence-matching task, an impairment that was quantitatively similar to that of the left-brain–damaged subjects.

A qualitative analysis yields some interesting explanations for this unexpected finding. Indeed, one of the characteristics of right-brain–damaged subjects is their tendency to select responses expressing an emotion that is the exact opposite of the one conveyed by the target item. At first glance, these errors could be interpreted as an incapacity of right-brain–damaged subjects to extract the emotional characteristics of a picture or a sentence. However, if one pays close attention to the explanations given by the patients for their choices, it becomes obvious that the errors have nothing to do with the decoding of the emotional content itself.

Rather, the problem seems to be their tendency to establish unexpected links between the pictures and the sentences. They tend to confabulate, to elaborate a new interpretation. For example, a patient might match the picture of an individual threatened with a gun with that of an individual receiving money, if he believed that the money was being forced upon the individual. Considering the instructions of the task, this response is wrong; but according to the subject's interpretation, the two scenes are similar from an emotional point of view. Cicone et al. (1980) suggested that it is not the emotional content itself that is the source of errors but that, generally speaking, the problem lies in the particular way in which these patients apprehend the entire linguistic or iconographic context involved.

On the basis of these findings, it would seem that the difficulties of right-brain–damaged subjects are not limited to a disorder of emotional content in verbal communication. This hypothesis has since been confirmed in a number of studies. For example, regardless of the information content, right-brain–damaged subjects exhibit impairments with respect to the organization or the comprehension of narrative speech, whether it involves recalling a story (Gardner et al., 1983; Wapner, Hamby, & Gardner, 1981) or organizing several sentences into a coherent story (Delis, Wapner, Gardner, & Moses, 1983; Huber & Gleber, 1982). The technique employed by Wapner et al. (1981) consisted in adding information to the flow of a story. This information could be either emotional or spatial in nature and could be either unusual or contradictory with respect to the main topic. After recalling the story, right-brain–damaged subjects ($N = 15$) had to answer a series of questions concerning this additional information. The results revealed no difference in the quality of recall as a function of whether the information was emotional or not. Furthermore, the quality of the answers given by the right-brain–damaged subjects to the emotional information was similar to the findings of Cicone et al. (1980): when the identified emotion is not as expected, it is nonetheless logical and adapted given the reinterpretation of the whole story provided by the subject.

This question aside, much of the Wapner et al. (1981) paper is devoted to a presentation of the supposedly different behavioral characteristics of right-brain–damaged subjects ($N = 10$). According to these authors, what distinguishes these patients from aphasics is their obvious tendency to make comments or to deviate from the topic of a story. Story recall produces frequent digressions of a personal nature as well as criticisms with respect to the content of the story. Sometimes the subjects provide their own conclusions or elaborate an entirely new story on the basis of one specific detail in the original story ("tangential speech"). Wapner et al. also emphasize the somewhat particular attitude of right-brain–damaged subjects with respect to unusual or contradictory information. Indeed, right-brain–damaged subjects are able to recall in detail these unusual elements and even have the tendency to justify them.[4] This attitude is in sharp

[4]One example of this attitude comes from the story of a boss who scolds an employee who is caught sleeping on the job. The incongruous sentence is "He offered him a salary increase." A right-brain–damaged subject who is asked to recall this sentence would justify it by a statement of the type: "The cost of living is increasing."

contrast with that of normals ($N = 10$) or aphasics ($N = 10$), who voluntarily set these contradictory statements aside or replace them in order to maintain the coherence of the whole story.

Right-brain–damaged subjects therefore seem capable of picking out these unusual incongruent elements but are incapable of rejecting them like normals or aphasics do. According to Wapner et al. (1981), one of the characteristics of right-brain–damaged subjects lies in this incapacity to judge the appropriateness of a given piece of information in reference to a specific context. Right-brain–damaged subjects seem to have problems with evaluating the plausibility of an event within a given context ("plausibility metrics"). Another characteristic of these subjects, still according to Wapner et al., concerns their difficulties interpreting a series of facts such as those contained in a fable. When questioned about the motivations behind the actions of the characters, right-brain–damaged patients respond correctly when these motivations are explicitly described in the text. Conversely, they perform poorly when these motivations are implicit, in other words, when they have to be inferred from the context.

In another paper (Gardner et al., 1983), the same authors described the narrative behavior of right-brain–damaged subjects in comparison to the performance of nine elderly subjects (65 to 85 years). The subjects were instructed to recall the major events of a fable. The performance of the right-brain–damaged subjects was significantly inferior to the performance of a group of normal subjects ($N = 10$) of identical age but similar to that of a group of elderly subjects. Likewise, deviant behaviors such as unnecessary comments, digressions, or confabulations were observed more frequently in right-brain–damaged subjects than in normals of the same age, whereas they were found just as often in the narrations of elderly subjects. Finally, half of the right-brain–damaged subjects as well as half of the elderly subjects were unable to find the moral of the fable, whereas 70% of the normal subjects were able to do so. This similarity between elderly subjects and right-brain–damaged subjects will be reexamined later in this chapter. One possible criticism of the studies conducted by Wechsler (1973), Wapner et al. (1981), and Gardner et al. (1983) concerns a possible memory overload, which might have been sufficient to interfere with the capacity to understand or to organize the story coherently.

However, other studies (Delis et al., 1983; Huber & Gleber, 1982) have shown that impaired discourse organization can exist independently of any gross memory disorder. In the Huber and Gleber study,[5] right-brain–damaged subjects ($N = 18$) were significantly impaired when they were required to put six sentences into a storylike sequence; their performance was in fact similar to a group

[5]These authors have also shown that right-brain–damaged patients perform poorly in similar tasks involving pictures instead of verbal material. However, their failure in this modality is somewhat difficult to interpret due to the methodology employed. Indeed, there were no attempts to ensure that the details of each picture were taken into consideration, a step which is necessary in trying to establish relationships between the different pictures.

of aphasics (i.e., 18 Broca's and 18 anomics). Delis et al. used a similar procedure: the subjects were presented with five randomly ordered sentences from which they had to elaborate a story, the topic of which was given by an introductory sentence. Three types of stories were compared, depending on the nature of the information content: (1) spatial information (e.g., a story describing the trail of a cat as it climbed onto the trunk of a car, then onto the roof, before jumping off), (2) temporal information (e.g., a story describing the various events in a doctor's workday), and (3) categorical information (e.g., a story describing the different stages in training a dog). According to the results, right-brain–damaged subjects (N = 10) were unable to correctly organize these stories in 50% of the cases, whereas normal control subjects (N = 10) were able to do so in 82% of the cases. Success rates also varied across the different types of stories: temporal information, 77%; spatial information, 44%; categorical information, 22%. Thus it seems that organizing sentences into a coherent sequence is a source of difficulty for right-brain–damaged patients, whatever the nature of the information conveyed.

The significant differences between the three types of stories warrant further explanation than that offered by Delis et al. (1983). Indeed, on the basis of the examples given by the authors, each story type requires operations of a different nature. For example, stories with temporal information are easy to elaborate since the subject needs only to rely on lexicon-based knowledge. Indeed, the simple presence of the words "morning," "noon," "afternoon," and "dinner" allows the sentences to be organized coherently. In stories containing spatial information, the subjects have to consider the whole sentence and then relate the various details provided by the different prepositions of place (e.g., above, on). Finally, stories containing categorical information are fundamentally different from the other two types since they require that subjects not content themselves with the information provided but rather that they interpret the information. For example, the sentence "The dog accepted the caress" means a greater familiarity with the dog than the sentence "The dog is beginning to stay in the same room as its master," but these differences are not explicit. The difference between these two sentences requires that the reader interpret the contextual information. The poor performance of right-brain–damaged subjects on this latter task should be paralleled with their difficulty in extracting implicit meaning (e.g., the motivations of the characters in a fable, when these are not explicitly given; Gardner et al., 1983; Wapner et al., 1981).

The studies of Huber and Gleber (1982) and Delis et al. (1983) have provided further understanding of the microstructural aspects of discourse organization. Text coherence is based on various linguistic elements that serve to relate the sentences between themselves (e.g., pronoun reiteration or lexical substitution). It is therefore legitimate to ask if the degree of redundancy of these cohesion marks is likely to help right-brain–damaged subjects reorganize a series of sentences into a coherent whole. To examine this question, Huber and Gleber (1982) and Delis et al. (1983) administered two different versions of each of their stories. The two versions differed by the number of cohesion marks. In both studies, the performance of right-brain–damaged subjects was not changed by the degree of redundancy of the cohesion marks. Thus the difficulty of right-brain–damaged subjects to organize narrative material does not seem to be situated at the so-called surface level of discourse.

At this stage of our review, the descriptive aspects of the behavior of right-brain–damaged subjects with respect to the organization and the comprehension of narrative speech can be summarized as follows: (1) difficulties in synthesizing a set of information coherently (e.g., to elaborate or reconstruct a story); (2) difficulties in rejecting minimally plausible incongruous events by reference to a given context; and (3) difficulties in interpreting the information implicitly contained in the context (e.g., finding the moral of a fable). However, it is not yet known if these characteristics of right-brain–damaged discourse are *specific* to this population if these characteristics are to be found in all right-brain–damaged subjects. The following paragraphs address these issues.

As just mentioned, the first issue concerns the specificity of a discourse impairment in right-brain–damaged subjects. In other words, to what extent is this impairment due specifically to the presence of a lesion of the right hemisphere? The similarity between the performance of right-brain–damaged subjects and that of a group of elderly patients (Gardner et al., 1983)[6] suggests that the difficulties of right-brain–damaged subjects at the discourse level are not specific; they are likely to be observed in any subject who is not in full possession of his optimal capacities, whether this is due to aging or to a lesion of the right hemisphere. The inclusion of a control group of nonaphasic left-brain–damaged subjects could help differentiate between the effects of a nonspecific lesion effect and the effects of a right-hemisphere lesion, despite the fact that *nonaphasic* left-brain–damaged subjects would certainly constitute a biased population in terms of lesion localization. In fact, even if it were possible to identify a group of nonaphasic left-brain–damaged subjects with a lesion in the perisylvian area, this group would *also* constitute a biased population since these subjects could correspond to the reversed equivalent of crossed aphasia (i.e., right-handed patients with aphasia following a lesion of the right hemisphere).

The problem of the specificity of the effects of a right-hemisphere lesion on a particular aspect of verbal communication is not specific to discourse. Indeed, it has also been discussed with respect to lexicosemantic functioning (Gainotti et al., 1983; see Chapter 4) and prosody (Ryalls et al., 1987; see Chapter 6). Gardner et al. (1983) concluded, however, that the similarity between right-brain–damaged and elderly subjects was only superficial. Indeed, these two populations could be differentiated by way of qualitative analyses. For example, elderly subjects would attend more to the chronology of the story's theme. Likewise, they would be more critical of their productions than right-brain–damaged subjects. Finally, and contrary to right-brain–damaged subjects, they would not exhibit

---

[6]The other interesting aspect of this comparison between brain-damaged and elderly subjects is linked to the controversial hypothesis according to which the "functions of the right hemisphere" are the first to be affected in aging (Goldstein and Shelly, 1981).

the tendency to choose the literal explanations that were presented. There are two reasons for using the conditional tense to report these so-called qualitative differences: (1) the lack of statistical information to support the evidence and (2) the authors did not rely on a specific model of storytelling in their descriptions.

The second issue concerns the characteristics of a sample population of right-brain–damaged subjects: authors usually report performance as if they were talking about a homogeneous population. In fact, we do not know if communication disorders affect each patient to the same degree. The clinical correlate of this question is that it is not known if these disorders are related to an actual communication disorder, such as the one observed in a first and unique interview. This raises the question of whether each of the surface-level characteristics of discourse impairment described earlier corresponds to the disturbance of a specific set of cognitive processes or if a more fundamental impairment of social and/or cognitive functioning is at the origin of these various communication disorders. A study conducted by Joanette et al. (1986) will allow us to examine these issues concerning the interpretation of speech disorders in right-brain–damaged subjects.

The first objective of Joanette et al. (1986; Joanette & Goulet, 1986b, 1990) was to revise the preceding descriptions using a model-based text analysis. In this study, the narratiave speech of 36 right-brain–damaged subjects was compared with that of 20 normal controls. The narrations were based on a series of eight pictures. In contrast to earlier studies (Cicone et al., 1980; Gardner et al., 1983; Wapner et al., 1981; Wechsler, 1973), the subjects' comments concerning their performance—the "modalizing" component (Nespoulous, 1980)—were separated from the narration itself—the "referential" component. Only this latter component was included in the analyses. Three different types of analyses (formal aspects, coherence, and informative content) were used to describe the referential component.

The formal aspect of the narrations was evaluated by way of classical indices such as the percentages of a given class of words (e.g., verb/noun or adjective/noun ratios) as well as descriptions of syntactic structure (e.g., the number of relative clauses). This analysis did not reveal any significant difference between the two groups of subjects except for a reduced amount of qualifications as revealed by the smaller percentages of adjectives and the higher noun/adjective ratio in the narrations of right-brain–damaged subjects.

The analysis of coherence was based on a model proposed by Charolles (1976, 1978), according to which a text should respect four metarules in order to be coherent:

1. The rule of *repetition* refers to those aspects that ensure coreference (e.g., pronominalization). This rule overlaps the notion of cohesion that was discussed in reference to the Huber and Gleber (1982) and Delis et al. (1983) studies.
2. The rule of *progression* corresponds to the semantic renewal that must accompany the development of a text; over a period of time, the text must be enriched with new information so as not to become circular.

3. The rule of *noncontradiction* states that in order to be coherent, a text should not contain information that is contradictory with other implicit or explicit information of the text, except in specific context (e.g., a joke, sarcasm).
4. The rule of *relation* requires that new information should be related, in some way, to previously given implicit or explicit information.

In light of these four metarules, Joanette and Goulet (1986b, 1990) described four types of coherence "errors." When each of these error types was considered individually, no significant difference was found between right-brain–damaged subjects and normal controls. However, when the four types of errors were summed, half[7] of the right-brain–damaged subjects differed significantly from the normals. Stated otherwise, the narrations of certain right-brain–damaged subjects could be described as incoherent, at least with respect to Charolle's conception of coherence.

The third type of analysis deals with informative content. This analysis is distinct from the preceding one, since, for example, a narration could be coherent and yet vary from the theme of the original story and not contain the essential information. The analysis of informative content used by Joanette et al. (1986) concerns the number and the nature of "propositions" contained in the narration of each subject. This approach is based on the description of a story's microstructure as proposed in the model of Kintsch and Van Dijk (1975; Van Dijk & Kintsch, 1983). A proposition is a unit of information consisting of a predicate and one or several arguments. Propositions can be either simple (e.g., an adjective qualifying an object verb) or complex (e.g., a proposition having another proposition as an argument). Right-brain–damaged subjects provide less information[8] than normals and the organization of this information is simpler despite the fact that narrations from both groups are equivalent in terms of the number of words. A comparison of right-brain–damaged and control subjects solely on the basis of the 32 pieces of information contained in the presumed core of the story shows, once again, that right-brain–damaged subjects provide significantly less information than normal controls. The differences between the two groups of subjects take on even greater importance since knowledge of the formal aspects of the lexicon and syntax is similar in both groups and comments (modalization) are not taken into account in these analyses.

The convergence between the results of these analyses and those from previously reported group of studies should be underlined. In this narration production task, coherence errors are compatible with what Wapner et al. (1981) referred to as the loss of the connecting line in narrative speech, or what Huber and Gleber (1982) as well as Delis et al. (1983) described as a difficulty with the coherent organization of a group of sentences. The weaker quality of the informa-

---

[7]This figure should be kept in mind.

[8]It should be mentioned that this information deficit equally affects pictures to the *right* of the patient *and* those to the left, thereby rejecting the eventuality that hemineglect is primarily responsible for the lower amount of information provided.

tion conveyed by the narration overlaps the difficulties in extracting the essential information from a story or iconographic material (cf. Gardner et al. 1983; Huber & Gleber, 1982), or what others (e.g., Gardner et al., 1983; Wapner et al., 1981) described under the term of interpretation difficulties, or inferential difficulties.

Let us now consider the specificity of these discourse impairments as well as the nature of the problem that might possibly be responsible for these changes. The rather large size of the population studied by Joanette et al. (1986; Joanette & Goulet, 1990) provided an opportunity to examine these questions by way of cluster analysis. The subjects (36 right-brain–damaged subjects and 20 normals) were grouped together on the basis of similarities in the content of their narrations (e.g., copresence or coabsence of a given set of propositions). This method resulted in the discrimination of two groups: the first group (Group I) included 31 subjects, of which 15 were normals; the second group (Group II) included 25 subjects, of which only 5 were normals. Thus the 36 brain-damaged subjects were evenly distributed across the two groups: 16 in Group I and 20 in Group II. The formal analysis revealed that the number of words used was greater in Group I; the propositional analysis revealed that Group I subjects produced narrations containing a greater amount of information. The missing information in Group II dealt essentially with the second part of the narration, namely, the complication, whereas the information associated with the setting and the resolution was similar in both groups. Group II included a significantly higher number of right-brain–damaged subjects (20/25). However, the presence of normal subjects within this group warrants reconsideration of the nature of the discourse-level problem in right-brain–damaged subjects.

Since normal subjects produced narrations that were similar to those of the majority of right-brain–damaged subjects, it is difficult to imagine that this particular type of discourse is the expression of an impairment involving cognitive processes that are specific to discourse. In fact, it seems more logical to assume that there are at least two ways, or two *styles*, to elaborate a narration from a sequence of pictures. The question then is to determine why a majority of right-brain–damaged subjects, as well as certain normals, adopt one style rather than the other. This issue has already been examined by Gardner et al. (1983). The similarity between the discourse abilities of certain normals and those of most right-brain–damaged subjects suggests that the characteristics of discourse in right-brain–damaged subjects are not all the exclusive expression of a lesion of the right hemisphere. It seems more appropriate to conclude that a right-hemisphere lesion is likely to restrict subjects to the use of a particular type of discourse, which can sometimes also be observed in normals or elderly subjects.

This point needs to be developed in more detail since, in our view, it is at the center of a debate involving two different conceptions of the description of discourse abilities in right-brain–damaged patients. If one adopts a point of view similar to the one underlying the descriptions of Wapner et al. (1981), the lack of information given by the subjects in Group II could be interpreted in terms of a difficulty in extracting pertinent information, or a difficulty in drawing inferences from the context. This interpretation is doubtful since normal subjects, though few in number, also failed to provide this information. It would mean, for example, that these normal controls are impaired with respect to inferential processes. It seems more logical, however, to assume that a modification of global pragmatic attitude in right-brain–damaged subjects is responsible for the adoption of one narrative style than the other. This interpretation in terms of a modification of global pragmatic attitude is in sharp contrast with the interpretation given in previous studies (e.g., Delis et al., 1983; Gardner et al., 1983; Wapner et al., 1981). Generally speaking, it seems that these studies favor a description of the elementary components of discourse, each of which is considered as a discrete cognitive process, without taking into account the overall characteristics of communication behavior in brain-damaged subjects. These studies give the impression that the different characteristics of discourse in right-brain–damaged subjects translate the disturbance or the inaccessibility of the various cognitive processes used to produce or to understand discourse.

In the following paragraphs, these questions will be discussed, first by addressing the problem of the homogeneity of groups of right-brain–damaged subjects and then by showing how important it is to develop adequate tools to assess what is referred to here as the overall pragmatic attitude of subjects. In limiting themselves to group studies, previous investigators have assumed that each of the different individuals within a given group have similar communication disorders. Clinical experience shows that not all right-brain–damaged patients present with a disorder of verbal communication. This issue has already been discussed with respect to prosody disorders (see Chapter 6). Results of Joanette et al. (1986; Joanette & Goulet, 1990) have provided additional evidence for narrative abilities (see above). It might be that certain subjects, included in studies such as those of Wapner et al. (1981) or Gardner et al. (1983), present an obvious communication disorder outside of any particular task. Moreover, these patients, who are not explicitly described as being critical of their difficulties, might be labeled as indifferent by their family, or even anosognosic in a neurological examination. If this is the case, then it is fundamental to know if these changes in narrative abilities are observed only in this type of patient. If they are, then it might be better to consider the possibility that a modification of pragmatic attitude is responsible for these changes in discourse, rather than a series of specific disorders, each of which is responsible for one particular aspect of discourse impairment. However, it is also possible that certain patients included in these studies did not have a manifest impairment of verbal communication and that only specific aspects of discourse were affected. In this case, the discourse characteristics described could be the consequence of more specific mechanisms, beyond those of a global modification of communication abilities. Be that as it may, these observations emphasize the fact that not all right-brain–damaged subjects are likely to present with a change of discourse abilities. Obviously, subsequent investigations will first have to identify those right-brain–damaged subjects with a verbal communication disorder before attempting to describe the exact nature of these disorders.

Now let us try and define what is meant by a global disorder of pragmatic attitude by taking the example of a task such as the recall of a narration (Gardner et al., 1983; Wapner et al., 1981). The task consists in reading a story to the subject, who is then instructed to recall it. From a pragmatic point of view, the task does not correspond to a real need of providing information. The subject has no new information to tell the examiner, who, by definition, knows the story. From a pragmatic point of view, the only information worthy of this name that the subject could communicate concerns his personal history, his illness, or even his point of view concerning the story content. What has been labeled as comments, tangential speech, or digressions (Gardner et al., 1983; Wapner et al., 1981) is valid only from the point of view of the task, not from the point of view of the exchange of information. The task (e.g., a narration) requires that the subject not behave as if he were in a situation of effective communication between two individuals, but instead that he limit himself to the role of a good subject.

We already know to what extent this notion of information effectively exchanged with the examiner is important in the quality of the communication of aphasic patients (Davis & Wilcox, 1981; Holland, 1977). The results of Joanette et al. (1986) and the behavior described by the preceding authors is compatible with an explanation of this nature. Right-brain–damaged patients, as well as certain normals, do not seem to be able to assume the role of a subject who must tell a story as best he can. The tendency of right-brain–damaged subjects to provide an interpretation which is not the one expected can be the consequence of this global attitude, which consists in not accepting the prescribed convention that one must adhere to the strict framework of the experiment. It is possible that this pragmatic attitude also explains the behavior of elderly subjects, as noted by Gardner et al. (1983). Of course, even if this were true, it might account for only part of what has been described; the possibility of a disturbance at the level of specific processes, such as inferential processes (see Gardner et al., 1983), cannot be excluded. We emphasize, however, that it would be easier to incriminate such processes if we had sufficient evidence to ensure that the "global pragmatic attitude" of a given subject is not itself disturbed.

At present, it seems that there is a gap between the importance given to the evaluation of specific cognitive processes involved in discourse abilities and the absence of adequate tools to evaluate the global pragmatic component. Such an evaluation could be undertaken in two different, albeit complementary ways. The development of questionnaires could lead to a description of the various parameters concerning the overall communication behavior of subjects, as well as a description of their more or less preserved capacities to engage in effective communication within a given context. These questionnaires could be used as a sort of assessment of a subject's handicaps with respect to the more common aspects of communication. Another method, inspired from studies of aphasics (e.g., Davis & Wilcox, 1981), would be to evaluate this pragmatic attitude experimentally. The general principle would be to compare a subject's performance on a task (e.g., narration) involving a modification of the communication context in which the task is conducted. Consider, for example, an experimental paradigm similar to that of the Promoting Aphasics Communicative Effectiveness (P.A.C.E.) (Davis & Wilcox, 1981) in which it would be possible to evaluate the subject's ability to take into account the amount of shared knowledge. A task could be devised in order to compare a narration produced in a situation very similar to that of effective communication, in which the subject tells a story that the examiner has never heard about (the pictures supporting the story could be provided by a third person), with a narration produced in the usual experiment or clinical situation in which it is the examiner who points to the pictures (Joanette & Goulet, 1985).

In summary, the semiological contribution of the different studies of narration behavior in right-brain–damaged subjects is important. The results of these studies suggest that the presence of a right-hemisphere lesion is likely to cause important difficulties with respect to the organization and/or the comprehension of discourse, usually narrative discourse. These results also suggest that, beyond the knowledge of the elementary aspects of language (linguistic functioning per se), the right-brain–damaged patient is impaired with respect to the apprehension of context, a condition that is necessary for efficient verbal communication. However, these studies do not provide any clear answer as to the nature of this disorder. We know that not all right-brain–damaged patients show discourse problems, though none of these studies has attributed these difficulties to a particular lesion site. We know that several aspects of discourse are concerned, but current knowledge of right-brain–damaged subjects does not allow us to attribute responsibility to a disorder of global pragmatic attitude or to a disturbance of more specific cognitive processes.

Sentence Interpretation Deficits

In this section, we review a number of studies concerning the sentence-level processing capacities of right-brain–damaged subjects, in particular at the receptive level. The preceding studies on discourse abilities provided evidence supporting both a preservation *and* an impairment of these processes. For example, Cicone et al. (1980) showed that right-brain–damaged subjects are capable of relating two sentences or two pictures, but that the resulting interpretation was unexpected in light of the task requirements. Similarly, when these same patients elaborated their own stories on the basis of a single detail, this behavior also reflected their inability to adhere to the task requirements, though it did reflect their preserved ability to interpret a specific detail (Gardner et al., 1983; Wapner et al., 1981). There are, however, several facts suggesting that this sentence-level interpretation capacity is limited. For example, right-brain–damaged subjects have been shown to accept incongruent information (Wapner et al., 1981), to have difficulties in explaining the motivations of the characters in a story (Gardner et al., 1983), or to be seemingly incapable of understanding the metaphoric meaning of idiomatic expressions (Hier & Kaplan, 1980; Weinstein,

1964). It has also been suggested that the more interpretation a task requires, the more difficulties right-brain–damaged subjects will have in relating sentences (Delis et al., 1983).

The studies reviewed in this section allow a more precise definition of the limited interpretation abilities of right-brain–damaged subjects as well as the situations in which these abilities seem to break down. Communication, whether it be reading a book or following a conversation, requires a continual establishing of relationships between the different pieces of information, which are not always given explicitly. The task of the reader or the speaker is to provide an interpretation on the basis of the preceding information (*deductive* inferences) and to anticipate the various implications of this interpretation for the rest of the story or the conversation (*inductive* inferences). These anticipations are then likely to be verified or rejected in such a way that in the end, all of the information taken in is interpreted as coherently and as closely as possible in accordance with the context in which it was presented.

To study these interpretation, or inferential, processes Brownell et al. (1986) asked their subjects to link two sentences by way of a single coherent interpretation. For example, the first sentence was of the type "Sally took a pen and a sheet of paper to meet a movie star," whereas the second sentence forced a backward inference of the type "The article concerned movie stars' views on nuclear energy." Relating these two sentences consists in deducing that "Sally is a reporter covering a story" (correct inference). If one were to take into account the first sentence only, a plausible inference would be "Sally is going autograph hunting." This inference becomes implausible after the second sentence has been read. Following the presentation of each pair of sentences such as these, the authors assessed the inferential processes of right-brain–damaged subjects by asking them to accept or reject each of two propositions: one proposition was the correct inference, the other was an incorrect inference based on only one of the two sentences. The incorrect inference was associated either with the first sentence (as in the example above) or with the second sentence. The responses to the correct inferences were similar to those given by control subjects, thereby suggesting that the right-brain–damaged subjects were capable of pure inferencing (Brownell et al., 1986). However, right-brain–damaged subjects also accepted incorrect inferences in 46% of the cases, whereas control subjects accepted them in only 16% of the cases.

The important point to be made here is that erroneous responses were significantly more frequent when the incorrect inference was associated with the first instead of the second sentence. It is as if right-brain–damaged subjects were initially capable of making a correct interpretation but were unable to question it after having read the second sentence. In their explanation of this somewhat rigid behavior of right-brain–damaged patients, Brownell et al. (1986) considered only the linguistic context. It seems important though to also examine the pragmatic attitude of the subjects in such a task. Indeed, it should be emphasized that the method employed by the authors seems to encourage acceptance of the initial interpretation: if the subjects accept this erroneous interpretation, it is because it was offered to them! In a certain way, they are given two occasions to make a mistake: the first occasion is when the sentence containing the plausible but incorrect interpretation is presented and the second occasion is when the incorrect interpretation is offered. Unfortunately, the experiment was conducted in such a way that it is not possible to predict what the spontaneous interpretation of the subject would have been after the presentation of the two sentences.

In addition, this task certainly would have benefited from a control of the relative plausibility of the interpretations given to the subjects. This notion of plausibility—referring to the pragmatic probability of occurrence of an event given the general knowledge of the world as well as the specific knowledge shared by two or more individuals—has already been pinpointed by Wapner et al. (1981) but, like Brownell et al. (1986), they did not analyze it in detail. It would be important to determine whether the incorrect inferences that were accepted by the right-brain–damaged subjects were correlated with their degree of plausibility. For example, it is possible that for a normal population, requesting an autograph is more plausible than writing an article.

The difficulty of right-brain–damaged subjects when the task requires the calling into question of an initial interpretation appears similar in nature to their behavior with respect to incongruent or contradictory details included in a narration. Right-brain–damaged subjects seem to be able to pick out the unusual or contradictory elements but are unable to reject them (Wapner et al., 1981). The results obtained by Tompkins and Mateer (1985) during the preliminary phase of their study on emotional prosody (see Chapter 6) could be interpreted in much the same way. In this study, subjects had to answer inferential questions concerning a text which consisted of two propositions and a conclusion. When the conclusion was congruent with the propositions, the answers of right-brain–damaged subjects ($N = 8$) were similar to those of normal controls. Conversely, when the conclusion was incongruent with the preceding context, their judgments were erroneous.

In a recent study, Joanette and Goulet (1987) examined the competence for "pure inferencing" (Brownell et al., 1986) in right-brain–damaged subjects ($N = 28$), left-brain–damaged nonaphasics ($N = 9$), and normal controls ($N = 13$). The subjects were asked to judge as true or false three kinds of statements relating to the specific content of a short narration that had been previously read aloud: (1) information that was *explicitly* present in the narrative, (2) information that was only *implicitly* present in the narration and which had to be inferred, and (3) information that was neither explicitly nor implicitly present in the narration but which was nonetheless *plausible*. The three groups of subjects obtained similar results, thereby confirming that right-brain–damaged subjects are capable of making inferences. The only other finding of some interest was the tendency of right-brain–damaged subjects to make "true" judgments for statements that were plausible but not mentioned in the story. In fact, when a correct interpretation is given (Brownell et al., 1986), or when the context is congruent (Joanette & Goulet, 1987; Tompkins & Mateer, 1985), right-brain–damaged subjects' responses are similar to those of normals. Therefore, right-brain–damaged subjects appear to have preserved inferential processes. However, difficulties do arise when the context, whether it be general, specific, or linked to task constraints, contains or favors several competing interpretations.

A number of studies (e.g., Caramazza et al., 1976; McDonald & Wales, 1986; Read, 1981) have focused on a particular type of inference—*logical* inferences—involved in problem solving, or syllogisms. For example, "*A* is larger than *B*, *B* is larger than *C*. Which is smallest?" (Read, 1981). In accordance with the spatial representation model of syllogism solving (De Soto, London, & Handel, 1965; Huttenlocher, 1968), Caramazza et al. (1976) suggested that the problems exhibited by right-brain–damaged subjects in some of their studies are due to their difficulty in forming a spatial representation of the relation between the elements of a problem. This suggestion is questionable for at least two reasons. On the one hand, it assumes that imagery processes are controlled essentially by the right hemisphere, a conception that is not supported by experimental findings (Paivio, 1986; Read, 1981). On the other hand, it does not take into consideration the linguistic processes necessary for solving these problems, such as the markedness of the comparative adjectives (Clark, 1971). It might be that these processes themselves are a source of difficulties for right-brain–damaged subjects.

Joanette and Goulet (1987) compared the performance of right-brain–damaged ($N = 31$) and control subjects ($N = 13$) on a syllogism-solving task of the three-term, two-relation type. To reduce memory load effects, the premises were presented both orally and visually. Fifty-six syllogisms were elaborated as well as controlled for spatial (anchorage) and linguistic (markedness and congruency) parameters and the order of presentation of the premises. The findings did not reveal any significant differences between the two groups. One interesting finding, though, was the tendency of right-brain–damaged subjects to be sensitive first to the spatial and then to the linguistic parameters. This is a further indication that elementary inferential processes of right-brain–damaged subjects are not impaired.

Understanding Humor

The difficulties of right-brain–damaged subjects to relate seemingly contradictory items of information explains the interest of several authors in the ability of these subjects to understand humor (Bihrle, Brownell, & Gardner, 1986; Brownell, Michel, Powelson, & Gardner, 1983). Indeed, more general investigations of humor point to the fact that an incongruency is necessary in order for a context to be perceived as humorous (Chapman & Foot, 1981; Lefort, 1986). Two opposing viewpoints have emerged from experimental studies of normal adults and children. The first is that the simple detection of incongruity is sufficient to appreciate the humorous value of a statement or of a drawing (Nerhardt, 1977). The second is that the appreciation of humor results from solving the problem raised by the incongruity (Suls, 1983). The latter viewpoint is based on a two-stage model, confirmed by a study of brain-damaged subjects (Brownell et al., 1983). The first stage is characterized by a surprise effect due to the presence of an incongruity that is in disagreement with the subject's anticipations. The second stage consists in solving this incongruity by reinterpreting all of the information. The pleasure one has with respect to the humorous nature of a joke depends on the individual's ability to solve the incongruity. The joke is all the more funny or subtle as the level of complexity of the problem to be solved is greater (Birhle et al., 1986; Lefort, 1986). Before being able to understand what makes a statement funny, one has to be able to uncover the incongruent element which signals that the statement is indeed a joke. Then it is up to the subject to decide if the joke is funny or not. If right-brain–damaged subjects are capable of picking out an incongruent element in a given context, as Wapner et al. (1981) suggest, then they should be able to reach the first stage, that is to say, understand the humorous nature of a context, whether it is presented in sentences or pictures.

The results of the experiment conducted by Brownell et al. (1983) tend to suggest that right-brain–damaged subjects do indeed have knowledge of this formal aspect of humorous statements. Two sentences were presented to the subjects. For example: "The neighborhood borrower approached Mr. Smith on Sunday afternoon and inquired: 'Say Smith, are you using your lawn mower this afternoon?' 'Yes, I am,' Smith replied warily." The subjects were asked to indicate which one of four possible solutions, when added to these two sentences, would give a humorous meaning to the story. For example: "The neighborhood borrower then answered: 'Fine, then you won't be needing your golf clubs, I'll just borrow them.'" This proposition is incongruent with respect to the anticipation of the listener who, like Mr. Smith, understood that the neighbor wanted to borrow the lawn mower. It forms the humorous unfolding of the story since this incongruity could be understood by reinterpreting the original intention of Mr. Smith's neighbor. The erroneous choices included an incongruent proposition (e.g., "You know, the grass is greener on the other side") that did not lead to a humorous conclusion. The two other propositions represented plain congruent complements of the original sentences and did not include any potential surprises for the listener (e.g., "Do you think I could use it when you're done?" and "Gee, if only I had money, I could buy my own."). Right-brain–damaged subjects (N = 12) found the correct solution in only 61% of the trials, whereas control subjects (N = 12) were successful on 81% of the trials. The interesting fact is that half of the erroneous choices of the right-brain–damaged subjects dealt with incongruent propositions. In other words, these patients seem to have a certain aptitude for understanding that incongruity is an essential part of humor. Their failure to understand humorous meaning seems to hold more from their incapacity to establish a coherent link between the incongruity and the context. In this study, humor is seen as a particular contextual situation, likely to reveal the difficulty of right-brain–damaged subjects in relating contradictory information. Apart from this cognitivist approach, a more direct observation of the understanding of humor in right-brain–damaged subjects lies in the psychoaffective dimensions of humor (Chapman & Foot, 1981; Freud, 1928/1981): laughter, smiling, and pleasure are by definition the reactions one would expect in a humorous context.

We have already discussed this emotional component as it relates to the overall communication behavior of right-brain–damaged subjects (Gainotti, 1972; Weinstein & Kahn, 1955), as well as to their capacities to process linguistic emotional information (Wapner et al., 1981; Wechsler, 1973), be it prosodic (see Chapter 6) or iconographic (Cicone et al., 1980). According to Gardner, Ling, Flamm, and Silverman (1975), the study of the reactions of right-brain–damaged subjects to humorous stimuli provides an additional approach to the description of this behavior. They compared the performance of a group of 19 right-brain–damaged subjects and 41 aphasics on a task that consisted in selecting which one of four drawings was the most humorous. Sometimes the humor concerned the drawing itself, while at other times it was the caption that was humorous. Furthermore, the four drawings could represent different scenes or they could differ from each other by only a single detail.

The performance of the right-brain–damaged subjects remained unchanged whether the humor was in the drawing or in the caption. As expected, aphasics performed poorly when comprehension of the caption was required. The analysis tends to suggest that the aphasics' errors were different from those of the right-brain–damaged subjects. Indeed, the drawings least understood by normals (N = 14) were also a source of difficulty for aphasics, whereas no such correlation was found for the right-brain–damaged subjects. Finally, the attitude of the right-brain–damaged subjects was very different from that of the aphasics. Examining the various reactions of their subjects, Gardner et al. (1975) found that most of the right-brain–damaged subjects were more indifferent than the aphasics, although two right-brain–damaged patients displayed excessive reactions to all of the drawings. Stated otherwise, the performance of right-brain–damaged subjects exposed to humorous material seems to reflect the indifference or the overwhelming euphoria that can sometimes be seen in the everyday communication abilities of these patients (Gainotti, 1972).

In spite of its interest, this study does not allow for any precise description of the nature of the subjects' impairment. Although the motivating factor behind this study was the emotional component of humor, there is nothing in the methodology that would allow for a judgment of the eventual influence of emotional content. Such a judgment would require the use of humorous drawings with controlled emotional content. Furthermore, the use of iconographic material raises certain problems of interpretation: in both groups of subjects, the difficulties were greater when the humor was based on a specific detail than when it was based on an entire scene. It is therefore possible that part of the difficulty lies in the extraction of relevant visual information.

Dagge and Hartje (1985) used a similar task, but the methodology allowed for a more specific interpretation of their results. Following a preliminary study with 18 normal subjects, Dagge and Hartje distinguished two types of humorous drawings: (1) so-called simple drawings, which could be directly perceived as such as in which the humor is based on a simple incongruent perception, and (2) so-called complex drawings, which required an interpretation in order to be understood. This "simple–complex" distinction is, in a certain way, an attempt to differentiate two stages, "incongruity–solving the incongruity," which were mentioned previously. Furthermore, to avoid any perceptual bias, the subjects were instructed to describe each of the drawings before giving their answer. This precautionary measure was deemed necessary since details are often neglected by patients. If we consider only the cases in which the drawings were described correctly, then it would seem that the simple drawings were understood in much the same way by right-brain–damaged subjects (N = 13), normals (N = 12), and aphasics (N = 13). With respect to the complex drawings, however, both the right-brain–damaged and aphasic subjects performed poorly in comparison to normals.

According to Dagge and Hartje (1985), the effect of the complexity of the drawings on the performance of the right-brain–damaged subjects supports the involvement of a disorder which is more of a cognitive than of an emotional nature. This conclusion, which is in agreement with the findings of Brownell et al. (1983), requires further discussion, particularly in light of the information provided by the authors concerning their population of right-brain–damaged subjects. Indeed, 8 of the 13 right-brain–damaged subjects were described as anosognosic or indifferent. It seems legitimate then to question the role that a different pragmatic attitude may have played along with a possible disturbance of the cognitive processes involved.

Another way of evaluating the understanding of humor in right-brain–damaged subjects is to ask them to quantify their appreciation of the humorous nature of a story on a scale from 0 to 5 (Gardner et al., 1983). In one experiment, half of the stories were humorous and the other half humorless. The latter were constructed from the former by replacing the punch line with a congruent proposition. Right-brain–damaged subjects were able to distinguish these two types of stories, thereby suggesting that they were capable of understanding the humorous nature of stories.

In a second experiment, Gardner et al. (1983) compared the evaluations given by the same subjects for four different types of humor: puns, tricks, puzzles, and foils.[9] In the right-brain–damaged subjects, the evaluation scores were lowest for the foils and highest for the puns. In fact, the right-brain–damaged subjects tended to give higher evaluation scores than the normals, regardless of whether the item were funny. This finding could reflect a global effect of the task itself on the choices made by these patients. Both Gardner et al. (1983) and Birhle et al. (1986) view these results as indicative of the preservation of a certain sensitivity of right-brain–damaged subjects to the formal aspects of humor. This interpretation also has the advantage of being congruent with the Brownell et al. (1983) study. In reality, these tasks are of a rather global nature and do not allow the separation of judgments based on the sole perception of the incongruity from judgments attesting to a true access to the comprehension of humorous content.

[9]*Puns*: "What happened to the girl who swallowed a spoon? She couldn't stir!" *Tricks*: "Why do birds fly south for the winter? It's too far to walk!" *Puzzles*: "What speaks in every langauge in the world but never went to school? An echo!" *Foils*: "Why do the clouds move? Because the wind pushes them!"

In summary, these studies of the understanding of humor lead to the observation that the emotional component of humor was not specifically evaluated. The eventual influence of distinct emotional contents (Cicone et al., 1980; Wapner et al., 1981) has yet to be determined. Apart from this limitation, the results of these studies reflect an adequate appreciation of the formal aspects of humor by right-brain–damaged subjects. The Brownell et al. (1983) study and, to a certain extent, that of Dagge and Hartje (1985) contribute further support for a possible impairment of the processes involved in the interpretation of humorous content (e.g., solving the incongruity). Further studies are needed to define the nature of this interpretation disorder since, beyond the humorous propositions studied by Brownell et al. (1983), humor is dependent on a very large number of basic cognitive processes. For example, puns involve the ability to recall two different meanings of a single word (polysemy). The capacity to understand this type of humor is therefore directly dependent on the lexicosemantic processing of words (Birhle et al., 1986; Brownell et al., 1984; see Chapter 4). The humorous effect could also be the result of a difference between the literal meaning of a statement and its metaphoric meaning. As we will see in a subsequent section, right-brain–damaged subjects tend to have difficulties with metaphoric meaning, independent of any humorous meaning.

Speech Acts

Theories of speech acts (Austin, 1970; Searle, 1969, 1982) are based on the simplistic observation that a verbal message is rarely limited to the transmission of raw, literal information; rather, it communicates an intent. The term "speech act" signifies the intention underlying any verbal message such as an order, a request, a promise, a wish, or an assertion. The intention of the speaker can be given explicitly by the linguistic form itself. This is the so-called direct speech act such as in the sentence "Close the window!" However, in everyday communication, this intentionality is far from being always as explicit. Most of the time, the speech act is indirect. Thus in a statement of the type "The window is open," the speaker does not limit himself to producing a literal assertion. Instead, his intent is either to reprimand someone or to suggest, albeit indirectly, that the listener close the window. Understanding an indirect speech act means being able to assimilate what was explicitly said (i.e., something about an open window), but above all being able to shift from this literal meaning in order to grasp the intent of the speaker within a given context.

The direct–indirect paradigm has been used to evaluate the difficulties of right-brain–damaged subjects to process contextual information. The studies reviewed here are all based on indirect requests, a type of speech act that has attracted the particular attention of psycholinguists. For example, Clark and Lucy (1975) suggested that an indirect request was first decoded automatically, depending on its literal meaning, then confronted with the context, which leads to a rejection of this first interpretation, in order to finally arrive at the unfolding of the real intention of the speaker. One study (Green & Boller, 1974) found that even severe aphasics are able to understand the intentions underlying direct speech acts. Likewise, though to a lesser degree, they are able to use contextual cues in order to understand the intention of certain indirect statements such as "I would like to know if you can stand up." Wilcox, Davis, and Leonard (1978) also concluded that aphasics have the competence to grasp the intentions underlying indirect requests: aphasics' judgments are much better than one would predict on the basis of their performance on verbal comprehension tests alone. The method employed by Wilcox, Davis, and Leonard consisted in presenting filmed sequences ($N = 40$) of a scene involving two speakers. The subjects had to judge if the response given by one of the protagonists was appropriate to the indirect request of the other.

Using the similar methodology, Hirst, Ledoux, and Stein (1984) compared the performance of five Broca's aphasics and five right-brain–damaged subjects. The indirect requests were all of the conventional type, such as "Could you pass the salt," which in a usual context calls for an action (i.e., passing the salt shaker), or "Can you play tennis?," which calls for a verbal response (e.g., "Yes, I can play"). It turns out that Broca's aphasics are able to understand these expressions in their conventional meaning when they call for an action to be performed by the listener. They correctly judge the appropriate or inappropriate nature of the action. Conversely, right-brain–damaged subjects tend to accept inadequate actions when a verbal response is required. Consider, for example, the question "Can you play tennis?" Right-brain–damaged subjects will accept a scene in which the listener responds by playing tennis in the living room where the protagonists were. However, when the request has a conventional meaning, in other words, when it demands an action, then the judgments of the right-brain–damaged subjects are correct. On the basis of their findings, Hirst et al. concluded that right-brain–damaged subjects have a knowledge of the conventional meaning of a request, but that they cannot judge if the request is appropriate or not within a given context.

It seems that the methodology used by these authors justifies a rewording of their interpretation. Indeed, since the particularlity of right-brain–damaged subjects lies in their acceptance of inadequate scenes, one needs to know if the scenes are inadequate only with respect to the request or if they are themselves incongruent. The example given by the authors suggest that the scenes are indeed incongruent (e.g., playing tennis in a living room). Thus it would seem that it is the capacity to judge the plausibility of a scene that is involved here. Therefore, a more appropriate interpretation would be that the conventional nature of an indirect request associated with a scene does not help right-brain–damaged subjects reject incongruent scenes.

In contrast to the preceding studies, Weylman, Brownell, and Gardner (1986) used only a linguistic context to evaluate the comprehension of indirect requests. Depending on the meaning of four introductory sentences, the meaning of the request could be either literal or indirect. When asked to decide the meaning of indirect requests, it seems that right-brain–damaged subjects ($N = 11$) are less sensitive than normals ($N = 10$) to the linguistic context. These results confirm the conclusions of a similar unpublished study (Heeschen & Reisches, 1979, reported in Foldi et al., 1983) which describes the difficulty of right-brain–damaged subjects to extract the underlying intention of a linguistic context.

Weylman et al. (1986) also took note of the fact that not all indirect requests have the same status. Certain requests such as "can you" are much more conventional than others such as "are you able to." This distinction is of some importance since, contrary to the Clark and Lucy (1975) model, there is some evidence suggesting that the meaning of conventional indirect requests, when they are conventional, is automatically processed without any preceding passage through the literal meaning (Gibbs, 1982, 1986), a situation that is somewhat similar to that of "frozen" metaphors, which are discussed in the next section. Weylman et al. showed that choices of right-brain–damaged subjects seem to be independent of this more or less conventional feature, whereas aphasics ($N = 5$) understand indirect requests even better when they are conventional. Thus these first studies, whether based on judgments of behavior (filmed sequences) or linguistic responses, reveal a difficulty of right-brain–damaged patients to understand indirect speech acts.

These studies need to be followed up in order to provide further descriptions of the different operations involved: (1) difficulty in interpreting the request itself depending on its more or less conventional nature (Weylman et al., 1986), (2) difficulty in interpreting the context (verbal or nonverbal) (Hirst et al., 1984), and (3) difficulty in relating the request and the context.

Metaphors

Metaphors, much like an indirect language act, sarcasm, or irony, translates an intention that is different from the initial literal meaning. The use of metaphors is common practice in everyday language, whether through marked expression (e.g., "He has a heavy heart") or apparently trite expression (e.g., "The window is still open" in the sense that "negotiations are still possible"), both of which can have a metaphoric meaning in a given context. This aspect of communication is worth testing in right-brain–damaged patients in order to evaluate their ability to use contextual information. It should be remembered here that right-brain–damaged patients have difficulties explaining idiomatic expressions (Weinstein, 1964) and proverbs (Hier & Kaplan, 1980). In one study (Winner & Gardner, 1977), we will see that this area has actually been examined very little in brain-damaged studies, which is in sharp contrast with the richness of the theoretical and experimental models issuing from psycholinguistics.

Winner and Gardner (1977) evaluated the oral comprehension of metaphoric expressions (e.g., "To have a heavy heart") by asking subjects to designate the corresponding picture. In this example, the correct picture depicted an individual crying. Another picture depicted the literal meaning of the expression (i.e., an individual stumbling under the weight of a heart he is carrying). Two distracting pictures, corresponding to each of the terms of the expression (the picture of a weight to illustrate the adjective "heavy"; the other picture, that of a heart), were also presented. The choices made by right-brain–damaged subjects ($N = 20$) were distributed as follows: 43% correct answers and 40% literal answers. This is significantly different from normal subjects ($N = 10$), 73% and 19% respectively, and left-brain–damaged subjects with anterior lesions ($N = 14$), 67% and 14. The choices of a group of left-brain–damaged subjects with posterior lesions ($N = 18$) were distributed as follows: correct, 46%; literal, 24%; and distractor, 30%. In a second phase of the study, the subjects were asked to give an oral explanation of the metaphoric expression. Paradoxically, right-brain–damaged subjects gave satisfactory explanations in 85% of the cases, a result that is in sharp contrast with their choice of literal representations for the same expressions. This dissociation therefore suggests that it is more the nature of the task than the understanding of the metaphoric meaning that determines the results.

To a certain extent, right-brain–damaged subjects are unable to reject the literal interpretation offered. This behavior is not dissimilar to that reported by Brownell et al. (1983): right-brain–damaged subjects are perplexed in the presence of a context containing two potential explanations. Winner and Gardner (1977) provided further evidence for the possible nonspecific nature of this impaired performance. Indeed, on the picture-designation task, a group of seven Alzheimer patients gave choices somewhat similar (45% correct and 44% literal) to those of the right-brain–damaged subjects. In other words, the designation of literal representations seems to attest to a particular attitude with respect to the task itself, an attitude that does not appear to be specific to right-brain–damaged subjects. These findings are compatible with the interpretation given above in terms of a disturbance of pragmatic attitude within the context of a given task. Although the findings of the Winner and Gardner study provide further evidence concerning the behavior of right-brain–damaged patients, its relevance with respect to the evaluation of metaphoric comprehension can be criticized for several reasons.

First the use of iconographic material is questionable. Insofar as pictures themselves may induce problems of interpretation in these patients, it is possible to attribute at least some of their errors to such problems, independently of any consideration of metaphoric expression. This question is inherent to any test that attempts to evaluate the comprehension of linguistic material by way of pictures. We do not know the extent to which the subjects take into consideration the plausible or implausible nature of a scene in accepting or rejecting one picture more than another. This problem is all the more critical since the different iconographic representations of the literal meaning of expressions used are not equivalent in terms of their plausibility. An experiment by Stachowiak, Huber, Poeck, and Kerschensteiner (1977) provides some interesting evidence in this respect. The investigators hypothesized that the proximity between a literal meaning and a metaphoric meaning can vary depending on the nature of the metaphoric expressions. This proximity should be taken into consideration in tasks using iconographic representations. For example, in the expression "They took off his shirt" (translated from German), the literal meaning (i.e., the partners of an unlucky player took off his shirt) is judged to be similar to its metaphoric meaning. However, the representation of an expression such as "He feels like a fish in the water" is judged to be quite different from its metaphoric meaning. Stachowiak et al. evaluated the comprehension of a text that described a situation (e.g., a game of cards in which one of the players had lost all of his money), which was then summarized within the text itself by way of a metaphoric expression. As in the Winner and Gardner (1977) study, the subjects had to select the corresponding picture from a series of pictures containing, among other things, the one representing the literal interpretation. It seems that right-brain–damaged subjects ($N = 19$) as well as aphasics ($N = 19$; Broca's, Wernicke's, amnesic, and global) choose three times as many pictures of literal representations when these are more similar to the metaphoric meaning than when they are not. This result suggests that idiomatic expressions do not all have the same status when their literal meanings are represented in pictures. The closer these expressions are to metaphoric meaning, the more plausible they become to normal or brain-damaged subjects. In brief, beyond the necessity to control for the perception of details (Dagge & Hartje, 1985), it seems that the use of pictures is pertinent only if one also controls their degree of plausibility.

Another problem with the investigations of metaphoric interpretation in brain-damaged populations is illustrated by the study of Winner and Gardner (1977): the nature of the metaphoric expression was not controlled. In the same way that it is important to distinguish indirect requests depending on their more or less conventional nature (Gibbs, 1986), psycholinguistics has taught us to recognize different types of metaphors. Without going into a theoretical discussion (Searle, 1982) or into a review of studies with normal subjects (e.g., Gibbs, 1985; Gibbs & Gonzales, 1985; Ortony, Schallert, Reynolds, & Antos, 1978), one can schematically distinguish at least two types of metaphoric expressions: (1) "dead" or "frozen" metaphors, which correspond to idiomatic expressions (e.g., "to have a heavy heart"), and (2) "alive" or "active" metaphors, which correspond to the result of a true metaphoric process (e.g., "the rocking chair of the faculty" in reference to a chairman who cannot decide what action to take). Only the comprehension of the second type of metaphor (active), not that of the first type (frozen), would require prior access to the literal meaning (Ortony et al., 1978). However, only frozen metaphors were used in the Winner and Gardner (1977) study. Thus the correct oral explanation given by the right-brain–damaged subjects is paradoxal only in appearance. In reality, to explain such idiomatic expressions out of context, it is not necessary to reject any prior literal meaning, for, in a certain way their metaphoric meaning that has become the literal meaning (Searle, 1982).

These idiomatic expressions, however, do not all have the same status within semantic memory. Their comprehension and their memorization seem to vary as a function of their degree of frozenness as well as their syntactic flexibility, that is, depending on whether or not they lose their meaning after a syntactic transformation (e.g.,from the present to the past tense) (Gibbs, 1985; Gibbs & Gonzales,

1985). It is therefore important to know if the performance of right-brain–damaged subjects on these idiomatic expressions is dependent on this difference of status in semantic memory.

In opposition to these closed metaphors are open metaphors, expressions created by the subjects (e.g., "it is a rock" in reference to a roast beef). As indicated earlier, active metaphors force a passage through literal meaning in order to understand the metaphoric meaning (Ortony et al., 1978; Searle, 1982). It would therefore be this type of metaphoric expression that should be used to evaluate the ability of right-brain–damaged subjects to adequately process contextual information.

After having read the preceding discussion of indirect speech acts and metaphors, one is led to suspect the presence of a gap between theoretical and experimental knowledge in these two areas, on the one hand, and the nature of the studies that have examined these aspects of communication in right-brain–damaged subjects, on the other hand. This gap is a result of the very recent nature of these theoretical bases; their applications in the near-future are foreseeable. For example, Weylman et al. (1986) took into consideration the distinction made by Gibbs (1982, 1985) between conventional and nonconventional indirect speech acts. Also, contrary to Winner and Gardner (1977) or Stachowiak et al. (1977), who did not have access to any references pertaining to the comprehension of metaphors by normal subjects, more specific ways of controlling the relevant variables are available and could be tested in brain-damaged subjects. In addition to these research perspectives, which are based on psycholinguistic models, it is important that future studies place greater emphasis on the pragmatic attitude of subjects engaged in a communication situation. Indeed, most studies—whether they involve the interpretation of speech, the comprehension of humor, indirect speech acts, or metaphors—are based on *responses* to questions put forward by the examiner. It would be interesting to see what behavior these subjects would adopt if they were able to respond freely. For example, one could imagine a task in which subjects have to provide the ending to a story containing a humorous or metaphoric expression. This type of procedure would allow a distinction between a disturbance of what has been here referred to as a global pragmatic attitude and a disturbance of specific cognitive processes engaged in the interpretation of metaphors or indirect speech acts.

Conclusion

In this chapter, we have provided enough evidence to suggest that the presence of a lesion of the right hemisphere can, in some cases, cause substantial changes in the communication behavior of an individual. While mastering the elementary aspects of language, right-brain–damaged patients seem to be impaired in the use of language in context, although *not all* of them seem to have such problems. One need only consider how often indirect speech acts, even humor, are used in everyday life in order to understand the extent to which these patients can be affected socially. In line with the increasing contribution of pragmatics during the 1980s, the years to come should provide even more detailed accounts of these disorders. Throughout this chapter, we have indicated some of the perspectives in this field. The recent development of models issued from the study of normal subjects leaves open their future application with respect to the study of communication in right-brain–damaged subjects. Finally, as has been the case for aphasia treatment, which has benefited from the contributions of pragmatics during the past decade (Davis & Wilcox, 1985; Holland, 1977), one can only hope that communication disorders of right-brain–damaged patients will eventually be the object of therapeutic implications, at least with the aim of helping these patients to cope better with their communication problems.

Conclusion

Much of the evidence that has been presented and discussed in this book does not support the classical viewpoint that verbal communication is under the exclusive control of the left hemisphere. On the one hand, there is no longer any reason to believe that the left hemisphere is solely responsible for all linguistic activities. On the other hand, studies reviewed here indicate that the integrity of the right hemisphere is necessary for those cognitive processes that allow for many of the components of verbal communication which are not necessarily covered by linguistic processes per se.

We have attempted to be critical with respect to all of the problems and limitations concerning an accurate description of the contribution of the right hemisphere to verbal communication in right-handed individuals. We have insisted deliberately on the inherent limitations of most of these studies, be it their methodological limitations (e.g., the samples studied, the techniques used, or the analyses performed) or their conceptual limitations (e.g., the prevailing theoretical framework of pragmatics). However, we did pinpoint a certain number of converging facts, issueing from studies that differed either with respect to the sample studied or with respect to the methodology employed; these facts tend to indicate that the contribution of the right hemisphere concerns, at the very least, lexicosemantic processes, the comprehension and/or the expression of prosody, and the apprehension and the processing of contextual information that is necessary for efficient verbal communication.

As indicated by the amount of space devoted to the issue, it is the contribution of the right hemisphere to lexical semantics that has inspired the largest number of studies. Generally speaking, there is more evidence favorable to a certain *potential* of the right hemisphere for the semantic processing of words than there is for its actual *contribution* to this processing. The results of these studies suggest that frequent, concrete, and imageable substantives are most likely to be processed by the right hemisphere. Other findings, in particular those from studies of right-brain–damaged patients, lead to the suggestion that, beyond this potential, the integrity of the right hemisphere is indeed necessary for efficient lexicosemantic functioning. However, the question remains as to whether the right hemisphere's contribution is direct, that is, strictly lexicosemantic in nature, or whether it is indirect, that is, attributable to its general contribution to intellectual functioning. If this contribution is indeed direct, it remains to be determined whether the integrity of the right hemisphere is also necessary to maintain actual semantic knowledge or use of this knowledge. In spite of the lack of answers to these questions, the principle of a right-hemisphere contribution to lexical semantics has received sufficient empirical support to warrant its use as a premise in a large number of studies looking at the role of the right hemisphere in aphasic left-brain damage in which the semiology implicates the right hemisphere (e.g., deep dyslexia, semantic paralexias).

As far as prosody is concerned, we have explained how the respective roles of the right and left hemispheres have been studied, up until now, as a function of the classical dissociation between linguistic prosody and emotional prosody. It seems that such a simplistic dissociation does not account for the role of the right hemisphere. It is highly probable that the right hemisphere is involved in emotional prosody as well as in linguistic prosody. It also seems that the processes involved in emotional prosody do not solely depend on the integrity of the right hemisphere. In fact, the prosodic features of verbal communication appear to be the result of a contribution of each of the two hemispheres. The actual involvement of each of the hemispheres probably depends on the relations that exist between the acoustic features and the linguistic characteristics of the verbal message. Only an approch that will refer not solely to a simple linguistic–emotional dichotomy but also to the kind of acoustic analysis needed to process incoming stimuli will likely contribute further to explaining the exact nature of the contribution of the two hemispheres.

We also examined how concepts issueing from recent theories of discourse abilities, as well as from other areas of pragmatics, have succeeded in resituating language within the more natural framework of human verbal communication. Indeed, the ability to communicate with others involves more than the use of an arbitrary system such as language. Efficient communication with others also requires a coherent and structured organization of speech based on a consideration of the context in which the exchange occurs. The integrity of the right hemisphere seems to be particularly important for the contextual appropriateness of communication behavior. A major portion of the changes that occur in the verbal communication behavior of right-brain–damaged patients tends to suggest a change in the attitude of these patients when facing a communication situation. We have underlined the fact that such a change in communication attitude might help explain both the changes that occur with respect to their discourse abilities and the performance of right-brain–damaged patients when submitted to certain specific tasks, such as telling a story.

We have stressed the various methodological problems that limit many of the studies in this particular field, but we have also underlined the perspectives offered by each type of approach. First, the facts that have emerged from the studies reviewed here have led us to emphasize the importance of a systematic study of (1) changes in the communicative attitude of right-brain–damaged patients during verbal communication and (2) the evaluation of the plausibility of communicative events, whether in natural communicative exchanges or in some formal conditions. Second, we have emphasized how the use of recent theoretical models could help systematize the description of certain aspects of verbal communication (e.g., the nonliteral aspects of communication). Finally, one thing that has also emerged from this review is the fact that the right-brain–damaged population that is used in order to appreciate the effective contribution of the right hemisphere to communicative abilities is far from being a homogeneous population. Indeed, those studies that have looked at the proportion of right-brain–damaged subjects actually presenting a particular impairment under study have usually reported that not all subjects show this impairment. Thus there is probably a problem in using one source of evidence defined only by reference to the hemisphere in which a lesion lies. No one, for instance, would try to understand the specific contribution of the left hemisphere to language by studying all subjects with a left-hemisphere lesion together, be they aphasic or not, and regardless of their type of aphasia. Thus one of the challenges of future studies whose goal is to look at the *effective* contribution of the right hemisphere to verbal communication will be to propose subject-selection criteria—either within the context of single-case or group studies—which would incorporate some specific expectations as to the presence of one, or more than one, particular communicative ability impairment. This is not a small challenge, but overcoming it will probably be the only means by which it will be possible to seriously hope for some answer to the question of the effective contribution of the right hemisphere to specific components of verbal communication.

We hope that this book will have convinced the reader that verbal communication can no longer be conceived as being exclusively under the control of the left hemisphere. Although in many respects the exact nature of the contribution of the right hemisphere is yet to be clearly specified, its existence can now be proposed with some certainty. The perspectives that have been suggested here may one day open the way to a more accurate description of the functional organization of the brain and the exact contribution of each of its two hemispheres to verbal communication. If this enterprise does succeed, then it will have contributed the bridge that has to be established between components of the functional architecture of verbal communication and one of the first described characteristics of the functional organization of the brain for cognitive functions, namely, its lateralization.

References

Alajouanine, T., & Lhermitte, F. (1965). Acquired aphasia in children. *Brain*, *88*, 653–662.

Alajouanine, T., Lhermitte, F., & Ducarne de Ribaucourt, B. (1960). Les alexies agnosiques et aphasiques. In Th. Alajouanine (Ed.), *Les grandes activités du lobe occiptal* (pp. 235–260). Paris: Masson.

Alajouanine, Th., Ombredane, A., & Durand, M. (1939). *Le syndrome de désintégration phonétique dans l'aphasie.* Paris: Masson.

Albert, M.L., & Sandson, J. (1986). Perseveration in aphasia. *Cortex*, *22*, 103–115.

Allport, D.A. (1977). On knowing the meaning of words we are unable to report: The effect of visual masking. In S. Dornic (Ed.), *Attention and performance* (Vol. 6) (pp. 505–534). New York: Academic Press.

Archibald, Y.M. & Wepman, J.M. (1968). Language disturbance and nonverbal cognitive performance in eight patients following injury to the right hemisphere. *Brain*, *91*, 117–130.

Austin, J.L. (1970). *Quand dire c'est faire.* Paris: Le Seuil.

Axelrod, S., Haryadi, T., & Leiber, L. (1977). Oral report of words and word approximations presented to the left or right visual field. *Brain and Language*, *4*, 550–557.

Babinski, J. (1914). Contribution à l'étude des troubles mentaux dans l'hémiplégie organique cérébrale (anosognosie). *Revue Neurologique*, *22*, 845–848.

Baillarger, J. (1865). De l'aphasie au point de vue psychologique. In J. Baillarger (Ed.), *Recherches sur les maladies mentales* (p. 584). Paris: Masson.

Barry, C. (1981). Hemispheric asymmetry in lexical access and phonological encoding. *Neuropsychologia*, *19*, 473–478.

Basser, L.S. (1962). Hemiplegia of early onset and the faculty of speech with special reference to the effects of hemispherectomy. *Brain*, *85*, 427–460.

Basso, A., Lecours, A.R. Moraschini, S. Vanier, M. (1985). Anatomoclinical correlations of the aphasias as defined through computerized tomography: Exceptions. *Brain and Language*, *26*, 201–229.

Baum, S.R., Daniloff, J.K., Daniloff, R., & Lewis, J. (1982). Sentence comprehension by Broca's aphasics: Effects of some suprasegmental variables. *Brain and Language*, *17*, 261–271.

Baxter, D.M., & Warrington, E.K. (1983). Neglect dysgraphia. *Journal of Neurology, Neurosurgery and Psychiatry*, *46*, 1073–1078.

Bear, M., (1983). La spécialisation hémisphérique et les fonctions émotionnelles chez l'homme. *Revue Neurologique*, *139*, 23–33.

Beaumont, J.G. (1982a). Studies with verbal stimuli. In J.G. Beaumont (Ed.), *Divided visual field studies of cerebral organization* (pp. 57–86). New York: Academic Press.

Beaumont, J.G. (1982b). The split-brain studies. In J.G. Beaumont (Ed.), *Divided visual field studies on cerebral organization* (pp. 217–232). New York: Academic Press.

Beaumont, J.G. (1983). Methods for studying cerebral hemisphere function. In A.W. Young (Ed.), *Functions of the right cerebral hemisphere* (pp. 113–146). London: Academic Press.

Behrens, S. (1986). The role of the cerebral right hemisphere in the production of linguistic prosody: An acoustic investigation. Paper presented at B.A.B.B.L.E., Niagara Falls, Ontario.

Behrens, S.J. (1985). The perception of stress and lateralization of prosody. *Brain and Language, 26*, 332–348.

Benowitz, L.I., Bear, D.M., Rosenthal, R., Mesulam, M.M., Zaidel, E., & Sperry, R.W. (1983). Hemispheric specialization in nonverbal communication. *Cortex, 19*, 5–11.

Bentin, S., & Gordon, W. (1979). Assessment of cognitive asymmetries in brain-damaged and normal subjects: Validation of a test battery. *Journal of Neurology, Neurosurgery and Psychiatry, 42*, 715–723.

Benton, A.L. (1968). Differential behavioral effects in frontal lobe disease. *Neuropsychologia, 6*, 53–60.

Bertelson, P. (1972). Listening from left to right versus right to left. *Perception, 1*, 161–165.

Bertelson, P. (1981). The nature of hemispheric specialization: Why should there be a single principle? [Commentary on *The nature of hemispheric specialization in man*, by J.L. Bradshaw and N.C. Nettleton]. *Behavioral and Brain Sciences, 4*, 63–64.

Bertelson, P. (1982). Lateral differences in normal man and lateralization of brain function. *International Journal of Psychology, 17*, 173–210.

Bever, T.G. (1975). Cerebral asymmetries in humans are due to the differentiation of two incompatible processes: Holistic and analytic. *Annals of the New York Academy of Sciences, 263*, 251–262.

Bihrle, A.M., Brownell, H.H., Gardner, H. (1986). Humor and the right hemisphere: A narrative perspective. Unpublished manuscript.

Birkett, P. (1977). Measures of laterality and theories of hemispheric process. *Neuropsychologia, 15*, 693–696.

Bishop, D., & Byng, S. (1984). Assessing semantic comprehension: Methodological considerations, and a new clinical test. *Cognitive Neuropsychology, 1*, 233–243.

Blumstein, S., & Cooper, W. (1974). Hemispheric processing of intonation contours. *Cortex, 10*, 146–158.

Blumstein, S., & Goodglass, H. (1972). The perception of stress as a semantic cue in aphasia. *Journal of Speech and Hearing Research, 15*, 800–806.

Boeglin, J. (1985). Performance d'écoute dichotique chez des enfants affectés ou non de troubles du langage. Unpublished Ph.D. Thesis, Université de Montréal.

Boeglin, J., Goulet, P., & Joanette, Y. (1988). The relationship between age and aphasia type. *Canadian Journal of Neurological Sciences, 15*, 231–232.

Boles, D.B. (1983). Dissociated imageability, concreteness, and familiarity in lateralized word recognition. *Memory and Cognition, 11*, 511–519.

Boller, F. (1968). Latent aphasia: Right and left "non aphasic" brain-damaged patients compared. *Cortex, 4*, 245–256.

Boller, F., & Vignolo, L.A. (1966). Latent sensory aphasia in hemisphere-damaged patients: An experimental study with the Token test. *Brain, 89*, 815–830.

Bolter, J.F., Long, C.J., & Wagner, M. (1983). The utility of the Thurstone word fluency test in identifying cortical damage. *Clinical Neuropsychology, 5*, 77–82.

Borkowsky, J.G., Benton, A.L., & Spreen, O. (1967). Word fluency and brain damage. *Neuropsychologia, 5*, 135–140.

Bornstein, R.A. (1986). Contribution of various neuropsychological measures to detection of frontal lobe impairment. *International Journal of Clinical Neuropsychology, 8*, 18–22.

Bouma, H. (1973). Visual interference in the parafoveal recognition of initial and final letters of words. *Vision Research, 13*, 767–782.

Bowers, D., Coslett, H.B., Baven, R.M., Spreedie, L.J., & Heilman, K.M. (1987). Comprehension of emotional prosody following unilateral lesions: Processing defect versus distraction defect. *Neuropsychology, 25*, 317–328.

Bradshaw, J.L. (1980). Right-hemisphere language: Familial and nonfamilial sinistrals, cognitive deficits and writing hand position in sinistrals, and concrete–abstract, imageable–nonimageable dimensions in word recognition. A review of interrelated issues. *Brain and Language, 10*, 172–188.

Bradshaw, J.L., Burden, V., & Nettleton, N.C. (1986). Dichotic and dichaptic techniques. *Neuropsychologia, 24*, 79–90.

Bradshaw, J.L., & Gates, A. (1978). Visual field differences in verbal tasks: Effects of task familiarity and sex of subject. *Brain and Language, 5*, 166–187.

Bradshaw, J.L., Hicks, R.E., & Rose, B. (1979). Lexical discrimination and letter-string identification in the two visual fields. *Brain and Language, 8*, 10–18.

Bradshaw, J.L., & Nettleton, N.C. (1983). *Human cerebral asymmetry.* Englewood Cliffs, N.J.: Prentice-Hall.

Brandeis, D., & Lehmann, D. (1986). Event-related potentials of the brain and cognitive processes: Approaches and applications. *Neuropsychologia, 24*, 151–168.

Broca, P. (1865). Sur la faculté du langage articulé. *Bulletin de la société d'Anthropologie, 6*, 337–393.

Brodal, A. (1973). Self-observations and neuro-anatomical considerations after a stroke (in the right hemisphere). *Brain, 96*, 675–694.

Brown, J.W. (1976). The neural organization of language: Aphasia and lateralization. *Brain and Language, 3*, 482–494.

Brown, J.W. (1982). Hierarchy and evolution in neurolinguistics. In M.A. Arbib, D. Caplan & J.C. Marshall (Eds.), *Neural models of language processes* (pp. 447–467). New York: Academic Press.

Brown, J.W., & Jaffe, J. (1975). Hypothesis on cerebral dominance. *Neuropsychologia, 13*, 107–110.

Brownell, H.H. (1988). Appreciation of metaphoric and connotative word meaning by brain-damaged patients. In C. Chiarello (Ed.), *Right hemisphere contributions to lexical semantics* (pp. 19–32). New York: Springer-Verlag.

Brownell, H.H., Bihrle, A.M., Potter, H.H., & Gardner, H. (1985). Appreciation of alternative word meanings by left and right brain damaged patients. Paper presented at the Academy of Aphasia, Pittsburgh.

Brownell, H.H., Michel, D., Powelson, J., & Gardner, H. (1983). Surprise but not coherence: Sensitivity to verbal humor in right-hemisphere patients. *Brain and Language, 18*, 20–27.

Brownell, H.H., Potter, H.H., Bihrle, A.M., & Gardner, H. (1986). Inference deficits in right brain-damaged patients. *Brain and Language, 27*, 310–321.

Brownell, H.H., Potter, H.H., & Michelow, D. (1984). Sensitivity to lexical denotation and connotation in brain-damaged patients: A double dissociation? *Brain and Language, 22*, 253–265.

Brownell, H.H., Simpson, T.L., Bihrle, A.M., Potter, H.H., & Gardner, H. (1988). *Appreciation of metaphoric alternative word meanings by left and right brain-damaged patients.* Manuscript submitted for publication.

Bruyer, R. (1983). *Le visage et l'expression faciale: Approche neuropsychologique.* Brussels: Mardaga.

Bruyer, R., & Strypstein, E. (1985). Concreteness and imageability in lateral differences in word perception. *Cahiers de psychologie cognitive, 5,* 111–125.

Bruyer, R., & Tuyumbu, B. (1980). Fluence verbale et lésions du cortex cérébral: Performances et types d'erreurs. *L'Encéphale, 6,* 287–297.

Bryden, M.P. (1982). *Laterality: Functional asymmetry in the intact brain.* New York: Academic Press.

Bryden, M.P., & Allard, F. (1976). Visual hemifield differences depend on typeface. *Brain and Language, 3,* 191–200.

Bryden, M.P., Munhall, K., & Allard, F. (1983). Attentional biases and the right ear effect in dichotic listening. *Brain and Language, 18,* 236–248.

Bub, D., & Lewine, J. (1988). Different modes of word recognition in the left and right visual field. *Brain and Language, 33,* 161–168.

Burgess, C., & Simpson, G.B. (1988). Cerebral hemispheric mechanisms in the retrieval of ambiguous word meanings. *Brain and Language, 33,* 86–103.

Burklund, C.W., & Smith, A. (1977). Language and the cerebral hemispheres. *Neurology, 27,* 627–633.

Caizergues, R.C. (1879). Notes pour servir à l'histoire de l'aphasie. *Montpellier Médical, 42,* 178–180.

Cambier, J., Elghozi, D., Signoret, J.L., & Henin, D. (1983). Contribution de l'hémisphère droit au langage des aphasiques. Disparition de ce langage après lésion droite. *Revue Neurologique, 139,* 55–63.

Cappa S.F., Papagno, C., & Vallar, G. (1987). Language and verbal memory in right-brain-damaged patients (RBD): A comparison with left-brain-damaged (LBD) without aphasia. *Journal of Clinical and Experimental Neuropsychology, 9,* 263.

Caramazza, A. (1986). On drawing inferences about the structure of normal cognitive systems from the analysis of patterns of impaired performance: The case for single-patient studies. *Brain and Cognition, 5,* 41–66.

Caramazza, A., Gordon, J., Zurif, E.B., & De Luca, D. (1976). Right-hemispheric damage and verbal problem solving behavior. *Brain and Language, 3,* 41–46.

Castro-Caldas, A., & Botelho, M.A.S. (1980). Dichotic listening in the recovery of aphasia after stroke. *Brain and Language, 10,* 145–151.

Cavalli, M., De Renzi, E., Faglioni, P., & Vitale, A. (1981). Impairment of right-brain damaged patients on a linguistic cognitive task. *Cortex, 17,* 545–556.

Chapey, R. (1981). *Language intervention strategies in adult aphasia.* Baltimore: Williams & Wilkins.

Chapman, A.J., & Foot, H.C. (1981). The psychology of humour. *Trends in Neurosciences, 4,* 4–9.

Charolles, M. (1976). Grammaire de texte. Théorie du discours. Narrativité. *Pratiques, 1,* 11–12.

Charolles, M. (1978). Introduction aux problèmes de la cohérence des textes. *Langue Française, 48,* 7–42.

Chiarello, C. (1983a). Semantic activation and semantic access in the left and right hemispheres. Paper presented at B.A.B.B.L.E., Niagara Falls, Ontario.

Chiarello, C. (1983b). Lexical access in the left and right hemispheres: Priming by sound, meaning, and orthography. Paper presented at the 11th Annual Meeting of the International Neuropsychological Society, Mexico.

Chiarello, C. (1985). Hemisphere dynamics in lexical access: Automatic and controlled priming. *Brain and Language, 26*, 146–172.

Chiarello, C. (1986). Abstract and concrete words: Differential asymmetry may be postlexical. Paper presented at the 9th European Conference of the International Neuropsychological Society, Veldhoven, Holland.

Chiarello, C. (1988a). Semantic priming in the intact brain: Separate roles for the right and left hemispheres? In C. Chiarello (Ed.), *Right hemisphere contributions to lexical semantics* (pp. 59–69). New York: Springer-Verlag.

Chiarello, C. (1988b). Evidence for lexical semantic processing in the normal right hemisphere. Paper presented at the 26th Annual Meeting of the Academy of Aphasia, Montreal (Canada).

Chiarello, C. (1989, March). On the conflicting evidence for right hemisphere language ability in normal and aphasic populations. Paper presented at B.A.B.B.L.E.-NET, Niagara Falls, N.Y.

Chiarello, C., & Church, K.L. (1986). Lexical judgments after right- or left-hemisphere injury. *Neuropsychologia, 24*, 623–630.

Chiarello, C., Church, K.L., & Hoyer, W.J. (1985). Automatic and controlled semantic priming: Accuracy, response bias, and aging. *Journal of Gerontology, 40*, 593–600.

Chiarello, C., Senehi, J., & Nuding, S. (1987). Semantic priming with abstract and concrete words: Differential asymmetry may be postlexical. *Brain and Language, 31*, 43–60.

Cicone, M., Wapner, W., & Gardner, H. (1980). Sensitivity to emotional expressions and situations in organic patients. *Cortex, 16*, 145–158.

Clark, H.H. (1971). More about "adjectives, comparatives, and syllogisms": A reply to Huttenlocher and Higgins. *Psychological Review, 78*, 505–514.

Clark, H., & Lucy, P. (1975). Understanding what is meant from what is said: A study in conversationally conveyed requests. *Journal of Verbal Learning and Verbal Behavior, 14*, 56–72.

Code, C. (1987). *Language aphasia and the right hemisphere.* Chichester: Wiley.

Cohen, G. (1973). Hemispheric differences in serial versus parallel processing. *Journal of Experimental Psychology, 97*, 349–356.

Cohen, G., & Freeman, R. (1978). Individual differences in reading strategies in relation to handedness and cerebral asymmetry. In J. Requin (Ed.), *Attention and performance* (Vol. 7, pp. 411–426). Hillsdale, N.J.: Lawrence Erlbaum Associates.

Collins, A.M., & Loftus, E.F. (1975). Spreading activation theory of semantic processing. *Psychological Review, 82*, 407–428.

Coltheart, M. (1980). Deep dyslexia: A right hemisphere hypothesis. In M. Coltheart, K. Patterson, & J.C. Marshall (Eds.), *Deep dyslexia* (pp. 326–380). London: Routledge and Kegan Paul.

Coltheart, M. (1983). The right hemisphere and disorders of reading. In A.W. Young (Ed.), *Functions of the right cerebral hemisphere* (pp. 172–201). London: Academic Press.

Coltheart, M., Patterson, K., & Marshall, J. (1987). *Deep dyslexia* (2nd ed.). London: Routledge and Kegan Paul.

Cooper, W.E., Soares, C., Nicol, J., Michelow, D., & Goloskie, S. (1984). Clausal intonation after unilateral brain damage. *Language and Speech, 27*, 17–24.

Cosmides, L. (1983). Invariances in the acoustic expression of emotion during speech. *Journal of Experimental Psychology: Human Perception and Behavior, 9*, 864–881.

Coughlan, A.K., & Warrington, E.K. (1978). Word-comprehension and word-retrieval in patients with localized cerebral lesions. *Brain, 101*, 163–185.

Critchley, M. (1962). Speech and speech-loss in relation to duality of the brain. In V.B. Mountcastle (Ed.), *Interhemispheric relations and cerebral dominance* (pp. 208–213). Baltimore: John Hopkins Press.

Crockett, H.G., & Estridge, N.M. (1951). Cerebral hemispherectomy. *Bulletin of the Los Angeles Neurological Society, 16*, 71–87.

Cummings, J.L., Benson, D.F., Walsh, M., & Levine, H.L. (1979). Left to right transfer of dominance: A case study. *Neurology, 29*, 1547–1550.

Cutting, J. (1978). Study of anosognosia. *Journal of Neurology, Neurosurgery, and Psychology, 41*, 548–555.

Czopf, D. (1979). The role of the non-dominant hemisphere in speech recovery in aphasia. *Aphasia Apraxia Agnosia, 2*, 27–33.

Dagge, M., & Hartje, W. (1985). Influence of contextual complexity on the processing of cartoons by patients with unilateral lesions. *Cortex, 21*, 607–616.

Damasio, A.R., & Damasio, H. (1983). The anatomic basis of pure alexia. *Neurology, 33*, 1573–1583.

Damasio, A.R., & Damasio, H. (1986). Hemianopsia, hemiachromatopsia and the mechanisms of alexia. *Cortex, 22*, 161–169.

Dandy, W.E. (1928). Removal of right cerebral hemisphere for certain tumors with hemiplegia. *Journal of the American Medical Association, 20*, 823–826.

Danly, M., & Shapiro, B.E. (1982). Speech prosody in Broca's aphasia. *Brain and Language, 16*, 171–190.

Davis, A.G., & Wilcox, J.M. (1981). Incorporating parameters of natural conversation in aphasia treatment. In R. Chapey (Ed.), *Language intervention strategies in adult aphasia* (pp. 169–193). Baltimore: Williams & Wilkins.

Davis, A.G., & Wilcox, J.M. (1985). *Adult aphasia rehabilitation: Applied pragmatics.* San Diego: College Hill Press.

Dax, M. (1865). Lésions de la moitié gauche de l'encéphale coïncidant avec l'oubli des signes de la pensée (Lu au Congrès méridional tenu à Montpellier en 1836). *Gazette Hebdomadaire de Médecine et de Chirurgie* (2nd series, Vol. 2, pp. 259–262).

Day, J. (1977). Right-hemisphere language processing in normal right-handers. *Journal of Experimental Psychology: Human Perception and Performance, 3*, 518–528.

Day, J. (1979). Visual half-life recognition as a function of syntactic class and imageability. *Neuropsychologia, 17*, 515–519.

De Partz, M.P. (1986). Re-education of a deep dyslexic patient: Rationale of the method and results. *Cognitive Neuropsychology, 3*, 149–177.

De Renzi, E., & Vignolo, L.A. (1962). The token-test: A sensitive test to detect receptive disturbances in aphasics. *Brain, 85*, 665–678.

De Soto, C., London, M., & Handel, S. (1965). Social reasoning and spatial paralogic. *Journal of Personality and Social Psychology, 2*, 513–521.

Dejerine, J. (1892). Contribution à l'étude anatomo-pathologique et clinique des différentes variétés de cécité verbale. *Comptes rendus des séances et mémoires de la Société de Biologie, 4*, 61–90.

Dejerine, J. (1914). *Sémiologie des affections du sytème nerveux.* Paris: Masson.

Delis, D.C., Wapner, W., Gardner, H., & Moses, J.A. (1983). The contribution of the right hemisphere to the organization of paragraphs. *Cortex, 19*, 43–50.

Deloche, G., Seron, X., Scius, G., & Segui, J. (1987). Right hemisphere language processing: Lateral difference with imageable and nonimageable ambiguous words. *Brain and Language, 30*, 197–205.

Demeurisse, G., & Capon, A. (1985). Does the right hemisphere contribute to language recovery in Broca's and Wernicke's aphasia? Paper presented at the 8th European Conference of the International Neuropsychological Society, Copenhagen.

Demeurisse, G., Verhas, M., & Capon, A. (1984). Resting CBF sequential study during recovery from aphasia due to ischemic stroke. *Neuropsychologia, 22*, 241–246.

Dennis, M., Lovett, M., & Wiegel-Crump, C.A. (1981). Written language acquisition after left or right hemidecortication in infancy. *Brain and Language, 12*, 54–91.

Denny-Brown, D., Meyer, J.S., & Horenstein, S. (1952). The significance of perceptual rivalry resulting from parietal lesion. *Brain, 75*, 433–471.

Diggs, C.C., & Basili, A.G. (1987). Verbal expression of right cerebrovascular accident patients: Convergent and divergent language. *Brain and Language, 30*, 130–146.

Dillon, R.F., & Smeck, R.R. (1983). *Individual differences in cognition* (Vol. 1). New York: Academic Press.

Dordain, M., Degos, J.D., & Dordain, G. (1971). Troubles de la voix dans les hémiplégies gauches. *Revue de Laryngologie, 3-4*, 178–187.

Drews, E. (1984). Organization of lexical knowledge in the left and right hemisphere. Paper presented at the 7th European Conference of the International Neuropsychological Society, Aachen.

Drews, E. (1987). Qualitatively different organizational structures of lexical knowledge in the left and right hemisphere. *Neuropsychologia, 25*, 419–427.

Dunn, L.M. (1965). *Expanded manual for the Peabody Picture Vocabulary Test* (Vol. 1). Circle Pines, Minnesota: American Guidance Service.

Dyer, F.N. (1973). Interference and facilitation for color naming with separate bilateral presentations of the word and color. *Journal of Experimental Psychology, 9*, 314–317.

Eisenson, J. (1959a). Language dysfunction associated with right brain damage. *American Speech and Hearing Association, 1*, 107.

Eisenson, J. (1959b). *Linguistic and intellectual modifications associated with right brain damaging.* Unpublished manuscript.

Eisenson, J. (1960). A second report on a study of modifications of language function associated with right cerebral damage. Paper presented at the Congress of the American Speech and Hearing Association, Los Angeles.

Eisenson, J. (1961). Linguistic and intellectual modification associated with right cerebral damage. Paper presented at the Congress of the American Speech and Hearing Association, Chicago.

Eisenson, J. (1962). Language and intellectual modifications associated with right cerebral damage. *Language and Speech, 5*, 49–53.

Eisenson, J. (1973). Right-brain damage and higher intellectual functions. In J. Eisenson (Ed.), *Adult aphasia* (pp. 38–41). Appleton-Century-Crofts, Prentice-Hall.

Ellis, A.W. (1984). Reading, writing and dyslexia: A cognitive analysis. London: Lawrence Erlbaum Associates.

Ellis, A.W., & Marshall, J.C. (1978). Semantic errors or statistical flukes? A note on Allport's "On knowing the meaning of words we are unable to report." *Quarterly Journal of Experimental Psychology, 30*, 569–575.

Ellis, A.W., & Young, A.W. (1981). Visual hemifield asymmetry for naming concrete nouns and verbs in children between seven and eleven years of age. *Cortex*, *17*, 617–624.

Ellis, H.D., & Shepherd, J.W. (1974). Recognition of abstract and concrete words presented in left and right visual fields. *Journal of Experimental Psychology, 103*, 1035–1036.

Emmorey, K. (1984). Linguistic prosodic abilities of left and right hemisphere brain damaged adults. Paper presented at the annual meeting of the Academy of Aphasia, Los Angeles.

Feyereisen, P., & Lannoy, J de. (1985). *Psychologie due geste.* Brussels: Mardaga.

Feyereisen, P., & Seron, X. (1982a). Nonverbal communication and aphasia: A review. I. Comprehension. *Brain and Language*, *16*, 191–212.

Feyereisen, P., & Seron, X. (1982b). Nonverbal communication and aphasia: A review. II. Production. *Brain and Language*, *16*, 213–236.

Foldi, N.S., Cicone, M., & Gardner, H. (1983). Pragmatic aspects of communication in brain-damaged patients. In S.J. Segalowitz (Ed.), *Language functions and brain organization* (pp. 51–86). New York: Academic Press.

Franzon, M., & Hugdahl, K. (1986). Visual half-field presentations of incongruent color words: Effects of gender and handedness. *Cortex*, *22*, 433–445.

French, L., Johnson, D., Brown, I.S., van Bergen, F. (1955). Hemispherectomy for control of intractable epileptic seizures. *Journal of Neurosurgery*, *12*, 154–164.

Freud, S. (1981). L'humour. In *Le mot d'esprit et ses rapports avec l'inconscient.* Paris: Gallimard. (Original work published 1928)

Fromkin, V.A. (1971). The nonanomalous nature of anomalous utterences. *Language*, *47*, 27–52.

Gagnon, J., Joanette, Y., Goulet, P., & Cardu, B. (1988). Automatic and controlled priming in right- and left-brain-damaged right-handers. Paper presented at 26th Annual Meeting of the Academy of Aphasia, Montreal.

Gainotti, G. (1972). Emotional behavior and hemispheric side of the lesion. *Cortex*, *8*, 41–55.

Gainotti, G. (1986). Emotions et spécialisation hémisphérique: vers de nouvelles stratégies de recherche. *Bulletin de Psychologie*, *39*, 923–930.

Gainotti, G., Caltagirone, C., & Miceli, G. (1979). Semantic disorders of auditory language comprehension in right-brain-damaged patients. *Journal of Psycholinguistic Research*, *8*, 13–20.

Gainotti, G., Caltagirone, C., & Miceli, G. (1983). Selective impairment of semantic-lexical discrimination in right-brain-damaged patients. In E. Perecman (Ed.), *Cognitive processing in the right hemisphere* (pp. 149–167). New York: Academic Press.

Gainotti, G., Caltagirone, C., Miceli, G., & Masullo, C. (1981). Selective semantic-lexical impairment of language comprehension in right-brain-damaged patients. *Brain and Language*, *13*, 201–211.

Gardner, H., Brownell, H.H., Wapner, W., & Michelow, D. (1983). Missing the point: The role of the right hemisphere in the processing of complex linguistic materials. In E. Perecman (Ed.), *Cognitive processing in the right hemisphere* (pp. 169–191). New York: Academic Press.

Gardner, H., & Denes, G. (1973). Connotative judgments by aphasic patients on a pictorial adaptation of the semantic differential. *Cortex*, *9*, 183–196.

Gardner, H., Ling, P.K., Flamm, L., & Silverman, J. (1975). Comprehension and appreciation of humour in brain-damaged patients. *Brain*, *98*, 399–412.

Gardner, H., Silverman, J., Wapner, W., & Zurif, E. (1978). The appreciation of antonymic contrasts in aphasia. *Brain and Language, 6*, 301–317.

Gazzaniga, M.S. (1967). The split brain in man. *Scientific American, 217*, 24–29.

Gazzaniga, M.S. (1970). *The bisected brain.* New York: Appleton-Century-Crofts.

Gazzaniga, M.S. (1971). Right hemisphere language. *Neuropsychologia, 9*, 479–482.

Gazzaniga, M.S. (1983a). Right hemisphere language following brain bisection. A 20-year perspective. *American Psychologist, 38*, 525–537.

Gazzaniga, M.S. (1983b). Reply to Levy and to Zaidel. *American Psychologist, 38*, 547–549.

Gazzaniga, M.S., & Baynes, K. (1986). Right hemisphere language: Insights into normal language mechanisms. Paper presented at the 66th Annual Meeting of the Association for Research in Nervous and Mental Diseases, New York.

Gazzaniga, M.S., & Hillyard, S.A. (1971). Language and speech capacity of the right hemisphere. *Neuropsychologia, 9*, 273–280.

Gazzaniga, M.S., & Ledoux, J.E. (1978). *The integrated mind.* New York: Plenum.

Gazzaniga, M.S., Ledoux, J.E., & Wilson, D.H. (1977). Language, praxis, and the right hemisphere: Clues to some mechanisms of consciousness. *Neurology, 27*, 1144–1147.

Gazzaniga, M.S., Smylie, C.S., & Baynes, K. (1984). Profiles of right hemisphere language and speech following brain bisection. *Brain and Language, 22*, 206–220.

Gazzaniga, M.S., & Sperry, R.W. (1967). Language after section of the cerebral commissures. *Brain, 90*, 131–148.

Geffen, G., & Quinn, K. (1984). Hemispheric specialization and ear advantages in processing speech. *Psychological Bulletin, 96*, 273–291.

Geschwind, N. (1965). Disconnection syndromes in animals and man. *Brain, 88*, 585–644.

Gibbs, R.W. (1982). A critical examination of the contribution of literal meaning to understanding non-literal discourse. *Text, 2*, 9–28.

Gibbs, R.W. (1985). On the process of understanding idioms. *Journal of Psycholinguistic Research, 14*, 465–472.

Gibbs, R.W. (1986). What makes some indirect speech act conventional? *Journal of Memory and Language, 25*, 181–196.

Gibbs, R.W., & Gonzales, G.P. (1985). Syntactic frozenness in processing and remembering idioms. *Cognition, 20*, 243–259.

Gill, K.M., & McKeever, W.F. (1974). Word length and exposure time effects on the recognition of bilaterally presented words. *Bulletin of the Psychonomic Society, 4*, 173–175.

Glushko, R.J. (1979). The organization and activation or orthographic knowledge in reading aloud. *Journal of Experimental Psychology: Human Perception and Performance, 5*, 674–691.

Goldberg, E., & Costa, L.D. (1981). Hemisphere differences in the acquisition and use of descriptive systems. *Brain and Language, 14*, 144–173.

Goldstein, G., & Shelly, C. (1981). Does the right hemisphere age more rapidly than the left? *Journal of Clinical Neuropsychology, 3*, 65–78.

Goldstein, K. (1948). *Language and language disturbances.* New York: Grune and Stratton.

Goodglass, H., & Calderon, M. (1977). Parallel processing of verbal and musical stimuli in right and left hemisphere. *Neuropsychologia, 15*, 397–407.

Goodglass, H., Graves, R., & Landis, T. (1980). Le Rôle de l'hémisphère droit dans la lecture. *Revue Neurologique, 136*, 669–673.

Goodglass, H., & Kaplan, E. (1972). The assessment of aphasia and related disorders. Philadelphia: Lea and Fibiger.

Gorelick, P.B., & Ross, E.D. (1987). The aprosodias: Further functional-anatomical evidence for the organization of affective language in the right hemisphere. *Journal of Neurology, Neurosurgery and Psychiatry, 50*, 553-560.

Gott, P.S. (1973). Language after dominant hemispherectomy. *Journal of Neurology, Neurosurgery and Psychiatry, 36*, 1082–1088.

Goulet, P., & Joanette, Y. (1987). Semantic clustering in right-brain-damaged right-handers. Paper presented at the 10th European Conference of the International Neuropsychological Society, Barcelona.

Goulet, P., & Joanette, Y. (1988). Semantic processing of abstract words in right brain-damaged patient. *Journal of Clinical and Experimental Neuropsychology, 10*, 312.

Goulet, P., Joanette, Y., Gagnon, J., & Sabourin, L. (1989). Semantics in right-brain-damaged right-handers. *Journal of Clinical and Experimental Neuropsychology, 11*, 353.

Gowers, W.R. (1887). *Lectures on the diagnosis of diseases on the brain.* London: Churchill.

Grant, S.R., & Dingwall, W.O. (1985). The role of the right hemisphere in processing linguistic prosody. Paper presented at the 13th Annual Meeting of the International Neuropsychological Society, San Diego.

Graves, R., Landis, T., & Goodglass, H. (1981). Laterality and sex differences for visual recognition of emotional and non-emotional words. *Neuropsychologia, 19*, 95–102.

Green, J., & Boller, F. (1974). Features of auditory comprehension in severely impaired aphasics. *Cortex, 10*, 133–145.

Green, J. (1984). Effects of intrahemispheric interference on reaction times to lateral stimuli. *Journal of Experimental Psychology: Human Perception and Performance, 10*, 292–306.

Gross, M.M. (1972). Hemispheric specialization for processing of visually presented verbal and spatial stimuli. *Perception and Psychophysics, 12*, 357–363.

Grossman, M. (1981). A bird is a bird: Making reference within and without superordinate categories. *Brain and Language, 12*, 313–331.

Habib, M., Ali-Chérif, A., & Poncet, M. (1987). Age-related changes in aphasia type and stroke location. *Brain and Language, 31*, 245–251.

Hannequin, D., Deloche, G., Branchereau, L., & Nespoulous, J.L. (1986). Noun/verb disambiguation in French and sentence comprehension in Broca's aphasics. Paper presented at the Annual Meeting of the Academy of Aphasia, Nashville.

Hannequin, D., Goulet, P., & Joanette, Y. (1987). *La contribution de l'hémisphère droit à la communication verbale.* Paris: Masson.

Harrington, A. (1985). Nineteenth-century ideas on hemisphere differences and "duality" of mind. *Behavioral and Brain Sciences, 8*, 617–660.

Hartje, W., Willmes, K., & Weniger, D. (1985). Is there parallel or independent hemispheric processing of intonational and phonetic components of dichotic speech stimuli? *Brain and Language, 24*, 83–99.

Hécaen, H. (1976). Acquired aphasia in children and the ontogenesis of hemispheric functional specialization. *Brain and Language, 3*, 114–134.

Hécaen, H., De Agostini, M., & Monzon-Montes, A. (1981). Cerebral organization in left-handers. *Brain and Language, 12*, 261–284.

Hécaen, H., & Dubois, J. (1969). *La naissance de la neuropsychologie du langage.* Paris: Flammarion.

Hécaen, H., & Marcie, P. (1974). Disorders of written language following right hemisphere lesions: Spatial dysgraphia. In S.J. Dimond & J.G. Beaumont (Eds.), *Hemisphere function in the human brain* (pp. 345–365). New York: Wiley.

Heilman, K.M., Bowers, D., Speedie, L., & Coslett, H.B. (1984). Comprehension of affective and nonaffective prosody. *Neurology, 34*, 917–921.

Heilman, K.M., Scholes, R., & Watson, R.T. (1975). Auditory affective agnosia. Disturbed comprehension of affective speech. *Journal of Neurology, Neurosurgery and Psychiatry, 38*, 69–72.

Heiss, W.D., Herholz, G., Pawlik, G., Wagner, R., & Wienhard, K. (1986). Positron emission tomography in neuropsychology. *Neuropsychologia, 24*, 141–149.

Hellige, J.B., & Cox, P.J. (1976). Effects of concurrent verbal memory on recognition of stimuli from left and right visual fields. *Journal of Experimental Psychology: Human Perception and Performance, 2*, 210–221.

Hellige, J.P., Cox, P.J., & Litvac, L. (1979). Information processing in the cerebral hemispheres: Selective hemispheric activation and capacity limitations. *Journal of Experimental Psychology, 108*, 251–279.

Hellige, J.B., & Sergent, J. (1986). Role of task factors in visual field asymmetries. *Brain and Cognition, 5*, 200–222.

Hellige, J.B., & Webster, R. (1979). Right-hemisphere superiority for initial stages of letter processing. *Neuropsychologia, 17*, 653–660.

Helm-Estabrooks, N. (1983). Exploiting the right hemisphere for language rehabilitation: Melodic intonation therapy. In E. Perecman (Ed.), *Cognitive processing in the right hemisphere* (pp. 229–240). New York: Academic Press.

Henderson, L. (Ed.) (1984). *Orthographies and reading: Perspectives from cognitive psychology, neuropsychology, and linguistics*. London: Lawrence Erlbaum Associates.

Henschen, S.E. (1926). On the function of the right hemispheres of the brain in relation to the left in speech music and calculation. *Brain, 49*, 110–123.

Hier, D.B., & Kaplan, J. (1980). Verbal comprehension deficits after right hemisphere damage. *Applied Psycholinguistics, 1*, 279–294.

Hines, D. (1976). Recognition of verbs, abstract nouns and concrete nouns from the left and right visual fields. *Neuropsychologia, 14*, 211–216.

Hines, D. (1977). Differences in tachistoscopic recognition between abstract and concrete words as a function of visual half-field and frequency. *Cortex, 13*, 66–73.

Hirst, W., Ledoux, J., & Stein, S. (1984). Constraints on the processing of indirect speech acts: Evidence from aphasiology. *Brain and Language, 23*, 26–33.

Hiscock, M., & MacKay, M. (1985). Neuropsychological approaches to the study of individual differences. In C. Reynolds & V. Willson (Eds.), *Methodological and statistical advances in the study of individual differences* (pp. 117–176). New York: Plenum.

Holland, A. (1977). Some practical considerations in aphasia rehabilitation. In M. Sullivan & M.S. Kommers (Eds.), *Rationale for adult aphasia therapy* (pp. 167–180). Lincoln: University of Nebraska Medical Center.

Howell, J.R., & Bryden, M.P. (1987). The effects of word orientation and imageability on visual half-field presentations with a lexical decision task. *Neuropsychologia, 25*, 527–538.

Huber, W., & Gleber, J. (1982). Linguistic and nonlinguistic processing of narratives in aphasia. *Brain and Language, 16*, 1–18.

Hugdahl, K., & Franzon, M. (1985). Visual half-field presentations of incongruent colorwords reveal mirror-reversal of language lateralization in dextral and sinistral subjects. *Cortex, 21*, 359–374.

Hughlings-Jackson, J. (1879). On affections of speech from disease of the brain. *Brain, 2*, 203–222.

Hughlings-Jackson, J. (1915). On the nature of the duality of the brain. *Brain, 38*, 80–103.

Huttenlocher, J. (1968). Constructing spatial images: A strategy in reasoning. *Psychological Review, 75*, 550–560.

Jeannerod, M., & Hécaen, H. (1979). *Adaptation et restauration des fonctions nerveuses.* Villeurbanne: Simep.

Joanette, Y. (1989). Aphasia in left-handers and crossed aphasia. In F. Boller & J. Grafman (Eds.), *Handbook of Neuropsychology* (Vol. 2, pp. 173–183). Amsterdam: Elsevier Science Publishers.

Joanette, Y., Ali-Chérif, A., Delpuech, F., Habib, M., Pélissier, J.F., & Poncet, M. (1983). Evolution de la sémiologie aphasique avec l'âge. Discussion à propos d'une observation anatomo-clinique. *Revue Neurologique, 139*, 657–664.

Joanette, Y., Drolet, M., & Goulet, P. (1989). Effects of right-hemisphere lesion on linguistic skills of right-handers: No change with age. Paper presented at the 17th Annual Meeting of the International Neuropsychological Society, Vancouver, British Columbia.

Joanette, Y., & Goulet, P. (1985). Contribution de l'hémisphère droit du droitier au langage et son évolution avec l'âge. Application for grant PG-28, C.M.R.C.

Joanette, Y., & Goulet, P. (1986a). Criterion-specific reduction of verbal fluency in right brain-damaged right-handers. *Neuropsychologia, 24*, 875–879.

Joanette, Y., & Goulet, P. (1986b). What narrative discourse tells us about the contribution of the right hemisphere to language. Paper presented at the Annual Meeting of the Academy of Aphasia, Nashville.

Joanette, Y., & Goulet, P. (1987). Inferencing deficit in right brain-damaged: Absence of evidence. Paper presented at the 10th European Conference of the International Neuropsychological Society, Barcelona.

Joanette, Y., & Goulet, P. (1988). Word-naming in right-brain-damaged subjects. In C. Chiarello (Ed.), *Lexical semantics and the right hemisphere* (pp. 1–18). New York: Springer-Verlag.

Joanette, Y., & Goulet, P. (1990). Narrative discourse in right-brain–damaged right-handers. In H.H. Brownell & Y. Joanette (Eds.), *Discourse ability and brain damage: Theoretical and empirical perspectives* (pp. 131–153). New York: Springer-Verlag.

Joanette, Y., Goulet, P., & Le Dorze, G. (1988). Impaired word naming in right-brain-damaged right-handers: Error types and time-course analysis. *Brain and Language, 34*, 54–64.

Joanette, Y., Goulet, P., Ska, B., & Nespoulous, J.L. (1986). Informative content of narrative discourse in right-brain damaged right-handers. *Brain and Language, 29*, 81–105.

Joanette, Y., Lecours, A.R., Lepage, Y., & Lamoureux, M. (1983). Language in right-handers with right-hemisphere lesions: A preliminary study including anatomical, genetic, and social factors. *Brain and Language, 20*, 217–248.

Johnson, J.P., Sommers, R.K., & Weidner, W.E. (1977). Dichotic ear preference in aphasia. *Journal of Speech and Hearing Research, 20*, 116–129.

Johnson, L.E. (1984). Vocal responses to left visual stimuli following forebrain commissurotomy. *Neuropsychologia, 22*, 153–166.

Jones, G.V. (1983). On double dissociation of function. *Neuropsychologia, 21*, 397–400.

Jones, G.V., & Martin, M. (1985). Deep dyslexia and the right-hemisphere hypothesis for semantic paralexia: A reply to Marshall and Patterson. *Neuropsychologia, 23*, 685–688.

Jonides, J. (1979). Left and right visual field superiority for letter classification. *Quarterly Journal of Experimental Psychology, 31*, 423–439.

Kay, J., & Marcel, T. (1981). One process, not two, in reading aloud: Lexical analogies do the work of non-lexical rules. *Quarterly Journal of Experimental Psychology, 33A*, 397–413.

Kertesz, A. (1982). *The Western aphasia battery.* New York: Grune and Stratton.

Kimura, D. (1961). Cerebral dominance on the perception of verbal stimuli. *Canadian Journal of Psychology, 15*, 166–171.

Kimura, D., & Harshman, R. (1984). Sex differences in brain organization for verbal and non-verbal functions. *Progress in Brain Research, 61*, 423–441.

Kinsbourne, M. (1971a). The minor hemisphere as a source of aphasic speech. *Archives of Neurology, 25*, 303–306.

Kinsbourne, M. (1971b). The minor hemisphere as a source of aphasic speech. *Transactions of the American Neurological Association, 96*, 141–145.

Kinsbourne, M., (1972). Behavioral analysis of the repetition deficit in conduction aphasia. *Neurology, 22*, 1126–1132.

Kinsbourne, M. (1975). The mechanisms of hemispheric control of the lateral gradient of attention. In P.M.A. Rabbit & S. Dornic (Eds.), *Attention and performance* (pp. 81–97). London: Academic Press.

Kinsbourne, M., & Warrington, E. (1962). A variety of reading disability associated with right hemispheric lesions. *Journal of Neurology, Neurosurgery and Psychiatry, 25*, 339.

Kintsch, W., & Van Dijk, T.A. (1975). Comment on se rappelle et on résume des histoires. *Language, 4*, 98–116.

Klix, F. (1978). On the representation of semantic information in human long-term memory. *Zeitschrift für Psychologie, 1*, 26–38.

Klonoff, H., & Kennedy, M. (1966). A comparative study of cognitive functioning in old age. *Journal of Gerontology, 21*, 239–243.

Kremin, H. (1980). Deux stratégies de lecture dissociables par la pathologie: Description d'un cas de dyslexie profonde et d'un cas de dyslexie de surface. *Grammatica, 7*, 131–156.

Kremin, H. (1984). Comments on pathological reading behavior due to lesions of the left hemisphere. In R.N. Malatesha & H.A. Whitaker (Eds.), *Dyslexia: A global issue* (pp. 273–309). The Hague: Matinus Nijhoff.

Laine, M. (1987). Correlates of word fluency performance. Paper presented at the 3rd Finnish Conference of Neurolinguistics. Johensuu, Finland.

Laine, M., & Niemi, J. (1988). Word fluency production strategies of neurological patterns. Semantic and phonological clustering. *Journal of Clinical and Experimental Neuropsychology, 10*, 28.

Lambert, A.J. (1982a). Right hemisphere language ability: 1. Clinical evidence. *Current Psychological Reviews, 2*, 77–94.

Lambert, A.J. (1982b). Right hemisphere language ability: 2. Evidence from normal subjects. *Current Psychological Reviews, 2*, 139–152.

Lambert, A.J., & Beaumont, J.G. (1981). Comparative processing of imageable nouns in the left and right visual fields. *Cortex, 17*, 411–418.

Lambert, A.J., & Beaumont, J.G. (1983). Imageability does not interact with visual field in lateral word recognition with oral report. *Brain and Language, 20*, 115–142.

Landis, T., Cummings, J.L., & Benson, D.F. (1980). Le passage de la dominance du langage à l'hémisphère droit: Une interprétation de la récupération tardive lors d'aphasies globales. *Revue Médecine Suisse Romande, 100*, 171–177.

Landis, T., Graves, R., & Goodglass, H. (1982). Aphasia reading and writing: possible evidence for right hemisphere participation. *Cortex, 18*, 105–112.

Landis, T., Regard, M., Graves, R., & Goodglass, H. (1983). Semantic paralexia: A release of right hemispheric function from left hemispheric control. *Neuropsychologia, 21*, 359-364.

Landis, T., Regard, M., & Serrat, A. (1980). Iconic reading in a case of alexia without agraphia caused by a brain tumour: A tachistoscopic study. *Brain and Language, 11*, 45–53.

Larmande, P., & Cambier, J. (1981). Influence de l'état d'activation hémisphérique sur le phénomène d'extinction sensitive chez dix patients atteints de lésions hémisphériques droites. *Revue Neurologique, 137*, 285–290.

Lebrun, Y., Lessinnes, A., De Vresse, L., & Leleux, C. (1985). Dysprosody and the non-dominant hemisphere. *Language Sciences, 7*, 41–52.

Lecours, A.R., Mehler, J., Parente, M.A., Caldeira, A., Cary, L., Castro, M.J., Dehaut, F., Delgado, R., Gurd, J., Fraga Karmann, D de., Jakubovitz, R., Osorio, Z., Cabral, L.S., & Junqueira, A.M.S. (1987). Illiteracy and brain damage. I. Aphasia testing in culturally contrasted populations (control subjects). *Neuropsychologia, 25*, 231–245.

Ledoux, J.E. (1983). Cerebral asymmetry and the integrated function of the brain. In A.W. Young (Ed.), *Functions of the right cerebral hemisphere* (pp. 203–216). London: Academic Press.

Ledoux, J.E. (1984). Cognitive function: Clues from brain asymmetry. In A. Ardila & F. Ostrosky-Solis (Eds.), *The right hemisphere: Neurology and neuropsychology* (pp. 51–59). London: Gordon and Breach.

Lee, H., Nakada, T., Deal, J.L., Lin, S., & Kwee, I.L. (1984). Transfer of language dominance. *Annals of Neurology, 15*, 304–307.

Lefort, B. (1986). Des problèmes pour rire. A propos de quelques approches cognitivistes de l'humour et de la drôlerie. *Bulletin de Psychologie, 40*, 183–195.

Lesser, R. (1974). Verbal comprehension in aphasia: An English version of three Italian tests. *Cortex, 10*, 247–263.

Levine, D.N., & Calvanio, R. (1982). Conduction aphasia. In H.S. Kirshner & F.R. Freemon (Eds.), *Neurolinguistics: Vol. 12. The neurology of aphasia* (pp. 79–111). Amsterdam: Swets & Zeitlinger B.V.

Levine, D.N., & Mohr, J.P. (1979). Language after bilateral cerebral infarctions: Role of the minor hemisphere in speech. *Neurology, 29*, 927–938.

Levy, J. (1978). Lateral differences in the human brain in cognition and behavioral control. In P.A. Buser & A. Rougel-Buser (Eds.), *Cerebral correlates of conscious experience* (pp. 285–298). Amsterdam: Elsevier–North-Holland Biomedical Press.

Levy, J. (1983a). Language, cognition and the right hemisphere. A response to Gazzaniga. *American Psychologist, 38*, 538–541.

Levy, J. (1983b). Is cerebral asymmetry of function a dynamic process? Implications for specifying degree of lateral differentiation. *Neuropsychologia, 21*, 3–11.

Levy, J. (1983c). Individual differences in cerebral hemisphere asymmetry: Theoretical issues and experimental considerations. In J.B. Hellige (Ed.), *Cerebral hemisphere asymmetry. method, theory and application* (pp. 465–497). New York: Praeger.

Levy, J., & Trevarthen, C. (1977). Perceptual, semantic and phonetic aspects of elementary language processes in split-brain patients. *Brain, 100*, 105–118.

Ley, R.G., & Bryden, M.P. (1982). A dissociation of right and left hemispheric effects for recognizing emotional tone and verbal content. *Brain and Cognition, 1*, 3–9.

Lifrak, M.D., & Novelly, R.A. (1984). Language deficits in patients with temporal lobectomy for complex-partial epilepsy. Paper presented at the 12th Annual Meeting of the International Neuropsychological Society, Houston.

Loftus, G.R. (1978). On interpretation of interactions. *Memory and Cognition*, *6*, 312–319.

Lupker, S.J., & Sanders, M. (1982). Visual-field differences in picture-word interference. *Brain and Cognition*, *1*, 381–398.

Mannhaupt, H.R. (1983). Processing of abstract and concrete nouns in a lateralized memory-search task. *Psychological Research*, *45*, 91–105.

Marcel, A.J., & Patterson, K.E. (1978). Word recognition and production: Reciprocity in clinical and normal research. In J. Requin (Ed.), *Attention and performance* (Vol. 7, pp. 209–226). Hillsdale, N.J.: Lawrence Erlbaum Associates.

Marcie, P., Hécaen, H., Dubois, J., & Angelergues, R. (1965). Les troubles de la réalisation de la parole au cours des lésions de l'hémisphère droit. *Neuropsychologia*, *3*, 217–247.

Marie, P., & Kattwinkel, (1897). Sur la fréquence des troubles du réflexe pharyngé et de la parole dans les lésions de l'hémisphere droit du cerveau. *Bulletin de la Société Médicale des Hopitaux*, pp. 516–518.

Marshall, J. (1981). Hemispheric specialization: What, how, and why? Commentary on "The nature of hemispheric specialization in man," by J.L. Bradshaw and N.C. Nettleton. *Behavioral and Brain Sciences*, *4*, 72–73.

Marshall, J. (1986). The description and interpretation of aphasic language disorder. *Neuropsychologia*, *24*, 5–24.

Marshall, J.C., & Patterson, K.E. (1983). Semantic paralexia and the wrong hemisphere: A note on Landis, Regard, Graves and Goodglass (1983). *Neuropsychologia*, *21*, 425–427.

Marshall, J.C., & Patterson, K.E. (1985). Left is still left for semantic paralexias: A reply to Jones and Martin (1985). *Neuropsychologia*, *23*, 689–690.

Martin, M. (1979). Hemispheric specialization for local and global processing. *Neuropsychologia*, *17*, 33–40.

Martin, R.C. (1982). The pseudohomophone effect: The role of visual similarity in nonword decisions. *Quarterly Journal of Experimental Psychology*, *34A*, 395–409.

McDonald, S., & Wales, R. (1986). An investigation of the ability to process inferences in language following right hemisphere brain damage. *Brain and Language*, *29*, 68–80.

McGlone, J. (1980). Sex differences in human brain asymmetry. *Behavioral and Brain Sciences*, *3*, 215–263.

McKeever, W.F., Sullivan, K.F., Ferguson, S.M., & Rayport, M. (1981). Typical cerebral hemisphere disconnection deficits following corpus callosum section despite spaning of the anterior commissure. *Neuropsychologia*, *19*, 745–755.

McKeever, W.F., Sullivan, K.F., Ferguson, S.M., & Rayport, M. (1982). Right hemisphere speech development in the anterior commissurotomy patient: A second case. *Clinical Neuropsychology*, *4*, 17–22.

McLeod, B.E., & Walley, R.E. (1988). Effects of ISI and target duration on priming interference. *Canadian Psychology*, *29*, 532.

McMullen, P.A., & Bryden, M.P. (1987). The effects of word imageability and frequency on hemispheric asymmetry in lexical disorders. *Brain and Language*, *31*, 11–25.

Meier, M.J., & Thompson, W.G. (1983). Methodological issues in clinical studies of right cerebral hemisphere dysfunction. In J.B. Hellige (Ed.), *Cerebral hemisphere asymmetry. Method, theory and application* (pp. 95–151). New York: Praeger.

Melville, J.P. (1957). Word length as a factor in differential recognition. *American Journal of Psychology, 70*, 316–318.

Miceli, G., Caltagirone, C., Gainotti, G., Masullo, C., & Silveri, M.C. (1981). Neuropsychological correlates of localized cerebral lesions in non-aphasic brain-damaged patients. *Journal of Clinical Neuropsychology, 3*, 53–63.

Milberg, W., & Blumstein, S.E. (1981). Lexical decision and aphasia: Evidence for semantic processing. *Brain and Language, 14*, 371–385.

Millar, J.M., & Whitaker, H.A. (1983). The right hemisphere's contribution to language: A review of the evidence from brain-damaged subjects. In S.J. Segalowitz (Ed.), *Language functions and brain organization* (pp. 87–113). New York: Academic Press.

Miller, L.K., & Butler, D. (1980). The effect of set size on hemifield asymmetries in letter recognition. *Brain and Language, 9*, 307–314.

Milner, B. (1962). Laterality effects in audition. In V. Mountcastle (Ed.), *Interhemispheric relations and cerebral dominance* (pp. 177–195). Baltimore: John Hopkins University Press.

Milner, B. (1964). Some effecs of frontal lobectomy in man. In M. Warren & K. Akert (Eds.), *The frontal granular cortese and behavior* (pp. 313–334). New York: McGraw-Hill.

Mirallié, C. (1896). *L'Aphasie sensorielle.* Unpublished doctoral dissertation, Faculté de Médecine de Paris.

Monrad-Krohn, G.H. (1947). Dysprosody or altered "melody of language." *Brain, 70*, 405–415.

Moore, W.H., & Weidner, W.E. (1975). Dichotic word-perception of aphasic and normal subjects. *Perceptual and Motor Skills, 40*, 379–386.

Morin, L., Joanette, Y., & Nespoulous, J.L. (1986). Grilles d'analyse des aspects pragmatiques de la communication inter-individuelle. *Rééducation Orthophonique, 24*, 137–149.

Moscovitch, M. (1976). On the representation of language in the right hemisphere of right-handed people. *Brain and Language, 3*, 47–71.

Moutier, F. (1908). *L'aphasie de Broca.* Paris: Steinheil.

Myers, J.J. (1984). Right hemisphere language: Science or fiction? *American Psychologist, 39*, 315–320.

Myers, J.J., & Sperry, R.W. (1985). Interhemispheric communication after section of the forebrain commissures. *Cortex, 21*, 249–260.

Neely, J.H. (1977). Semantic priming and retrieval from lexical memory: Roles of inhibition loss spreading activation and limited-capacity attention. *Journal of Experimental Psychology: General, 106*, 226–254.

Nerhardt, G. (1977). Operationalism of incongruity in humour research: A critique and suggestions. In J.A. Chapman & H. Foot (Eds.), *It's a funny thing, humour* (pp. 47–51). Oxford: Pergamon Press.

Nespoulous, J.L. (1980). Linguistique et aphasie. *Revue Neurologique, 136*, 637–650.

Nespoulous, J.L., Perron, P., & Lecours, A.R. (Eds.) (1986). *The biological foundations of gestures: Motor and semiotic aspects.* Hillsdale, N.J.: Lawrence Erlbaum Associates.

Newcombe, F. (1969). *Missile wounds of the brain.* London: Oxford University Press.

Newcombe, F., & Marshall, J.C. (1980). Response monitoring and response blocking in deep dyslexia. In M. Coltheart, K. Patterson, & J.C. Marshall (Eds.), *Deep dyslexia* (pp. 160–175). London: Routledge and Kegan Paul.

Newcombe, F., Oldfield, R.C., Ratcliff, G.G., & Windfield, A. (1971). Recognition and naming of object drawings by men with focal wounds. *Journal of Neurology, Neurosurgery and Psychiatry, 34*, 329–340.

Newcombe, F., Oldfield, R.C., & Windfield, A. (1965). Object-naming by dysphasic patients. *Nature*, *207*, 1217–1218.

Niccum, N. (1986). Longitudinal dichotic listening patterns for aphasic patients. I. Description of recovery curves. *Brain and Language*, *28*, 273–288.

Niccum, N., & Rubens, A.B. (1983). Reversal of ear advantage in recovery from aphasia. A case report. *Neuropsychologia*, *21*, 265–271.

Niccum, N., Selnes, O.A., Speaks, C., Risse, G.L., & Rubens, A.B. (1986). Longitudinal dichotic listening patterns for aphasic patients. III. Relationship to language and memory variables. *Brain and Language*, *28*, 303–317.

Niccum, N., Speaks, C., Rubens, A.B., Knopman, D.S., Yock, D., & Larson, D. (1986). Longitudinal dichotic listening patterns for aphasic patients. II. Relationship with lesion variables. *Brain and Language*, *28*, 289–302.

Nielsen, J.M. (1944). Function of the minor (usually right) cerebral hemisphere in language. *Bulletin of Los Angeles Neurological Society*, *2*, 67–75.

Oldfield, R.C. (1966a). Dénomination d'objets et stockage des mots. *Bulletin de Psychologie*, *247*, 733–744.

Oldfield, R.C. (1966b). Things, words and the brain. *Quarterly Journal of Experimental Psychology*, *18*, 340–353.

Ombredane, A. (1951). *L'Aphasie et l'élaboration de la pensée explicite.* Paris: Presses Universitaires de France.

Oreinstein, H.B., & Meighan, W.B. (1976). Recognition of bilaterally presented words varying in concreteness and frequency: Lateral dominance or sequential processing? *Bulletin of the Psychonomic Society*, *7*, 179–180.

Ortony, A., Schallert, D., Reynolds, R., & Antos, S. (1978). Interpreting metaphors and idioms: Some effects of context on comprehension. *Journal of Verbal Learning and Verbal Behavior*, *17*, 465–477.

Paivio, A. (1971). *Imagery and verbal process.* New York: Holt, Rinehart and Winston.

Paivio, A. (1986). *Mental representations: A dual coding approach.* Oxford: Oxford University Press.

Paivio, A., Yuille, J.C., & Madigan, S.A. (1968). Concreteness, imagery, and meaningfulness values for 925 nouns. *Journal of Experimental Psychology* (Monograph supplement) *76*, 1–25.

Paradis, M. (Ed.) (1983). *Readings on aphasia in bilinguals and polyglots.* Montréal: Didier.

Paradis, M., Hagiwara, H., & Hildebrandt, N. (1985). *Neurolinguistic aspects of Japanese writing system.* New York: Academic Press.

Parkin, A.J. (1984). Redefining the regularity effect. *Memory and Cognition*, *12*, 287–292.

Parkin, A.J., & Underwood, G. (1983). Orthographic versus phonological irregularity in lexical decision. *Memory and Cognition*, *11*, 351–355.

Parkin, A.J., & West, S. (1985). Effects of spelling-to-sound regularity on word identification following brief presentation in the right or left visual field. *Neuropsychologia*, *23*, 279–283.

Patterson, K.E. (1981). Neuropsychological approaches to the study of reading. *British Journal of Psychology*, *72*, 151–174.

Patterson, K.E., & Besner, D. (1984a). Is the right hemisphere literate? *Cognitive Neuropsychology*, *1*, 315–341.

Patterson, K.E., & Besner, D. (1984b). Reading from left: A reply to Rabinowicz and Moscovitch and to Zaidel and Schweiger. *Cognitive Neuropsychology*, *1*, 365–380.

Patterson, K.E., & Kay, J. (1982). Letter-by-letter reading: Psychological descriptions of a neurological syndrome. *Quarterly Journal of Experimental Psychology, 34*, 411–441.

Pendleton, M.G., Heaton, R.K., Lehman, R.A.W., & Hulihan, D. (1982). Diagnostic utility of the Thurstone word fluency test in neuropsychological evaluations. *Journal of Clinical Neuropsychology, 4*, 307–317.

Perecman, E. (1983). Introduction: Discovering buried treasure. A look at the cognitive potential of the right hemisphere. In E. Perecman (Ed.), *Cognitive processing in the right hemisphere* (pp. 1–16). New York: Academic Press.

Perecman, E., & Kellar, P. (1981). The effect of voice and place among aphasics, nonaphasic right-damaged, and normal subjects on a metalinguistic task. *Brain and Language, 12*, 213–223.

Peretz, I., Morais, J., & Bertelson, P. (1987). Shifting ear differences in melody recognition through strategy inducement. *Brain and Cognition, 6*, 202–215.

Perfetti, C.A. (1983). Individual differences in verbal processes. In R.F. Dillon & R.R. Smeck (Eds.), *Individual differences in cognition. Vol. 1* (pp. 65–104). New York: Academic Press.

Perret, E. (1974). The left frontal lobe of man and the suppression of habitual responses in verbal categorical behavior. *Neuropsychologia, 12*, 323–330.

Pettit, J.M., & Noll, J.D. (1979). Cerebral dominance in aphasia recovery. *Brain and Language, 7*, 191–200.

Poeck, K., de Bleser, R., & Graf von Keyserlingk, D. (1984). Computed tomography localization of standard aphasic syndromes. In F.C. Rose (Ed.), *Advances in neurology: Vol. 42. Progress in aphasiology* (pp. 71–89). New York: Raven Press.

Poncet, M., & Ponzio, J. (1975). Les troubles de la communication du langage chez des sujets du troisième âge. Paper presented at the Journées d'étude sur la pathologie du troisième âge, Marseille.

Porter, R.J., & Hughes, L.F. (1983). Dichotic listening to CVs: Method, interpretation and application. In J.B. Hellige (Ed.), *Cerebral hemisphere asymmetry. Method, theory and application* (pp. 177–218). New York: Praeger.

Posner, M.I., & Snyder, C.R.R. (1975). Facilitation and inhibition in the processing of signals. In P.M.A. Rabbit & S. Dornic (Eds.), *Attention and performance* (Vol. 5, pp. 669–682). New York: Academic Press.

Prinz, P. (1980). A note on requesting strategies in adult aphasics. *Journal of Communication Disorders, 13*, 65–73.

Rabinowicz, B., & Moscovitch, M. (1984). Right hemisphere literacy: A critique of some recent approaches. *Cognitive Neuropsychology, 1*, 343–350.

Ramier, A., & Hécaen, H. (1970). Rôle respectif des atteintes frontales et de la latéralisation lésionnelle dans les déficits de la "Fluence verbale." *Revue Neurologique, 123*, 17–22.

Rankin, J.W., Adam, D.M., & Horwitz, S.J. (1981). Language ability in right and left hemiplegic children. *Brain and Language, 14*, 292–306.

Read, D.E. (1981). Solving deductive reasoning problems after unilateral temporal lobectomy. *Brain and Language, 12*, 92–100.

Regard, H., & Landis, T. (1984). Experimentally induced semantic paralexias in normals: A property of the right hemisphere. *Cortex, 20*, 263–270.

Restatter, M., Dell, C.W., McGuire, R.A., & Loren, C. (1987). Vocal reaction times to unilaterally presented concrete and abstract words: Towards a theory of differential right hemispheric semantic processing. *Cortex, 23*, 135–142.

Rivers, D.L., & Love, R.J. (1980). Language performance on visual processing tasks in right hemisphere lesion cases. *Brain and Language, 10*, 348–366.

Rodel, M., Dudley, J.G., & Bourdeau, M. (1983). Hemispheric differences for semantically and phonologically primed nouns: A tachistoscopic study in normals. *Perception and Psychophysics*, *34*, 523–533.

Rodel, M., Landis, T., & Regard, M. (1989). Hemispheric dissociation in semantic relation. *Journal of Clinical and Experimental Neuropsychology*, *11*, 70.

Rosen, W.G. (1980). Verbal fluency in aging and dementia. *Journal of Clinical Neuropsychology*, *2*, 135–146.

Ross, E. (1981). The aprosodias: Functional-anatomical organization of the affective components of language in the right hemisphere. *Archives of Neurology*, *38*, 561–569.

Ross, E. (1984a). Disturbances of emotional language with right hemisphere lesions. In A. Ardila & F. Ostrosky-Solis (Eds.), *The right hemisphere: Neurology and neuropsychology* (pp. 109–123). London: Gordon and Breach.

Ross, E. (1984b). Right hemisphere's role in language, affective behavior and emotion. *Trends in Neuroscience*, *7*, 342–346.

Ross, E., Harney, J.H., de Lacoste-Utamsing, C., & Purdy, P.D. (1981). How the brain integrates affective and propositional language: A unified behavioral function. *Archives of Neurology*, *38*, 745–748.

Ross, E., & Mesulam, M. (1979). Dominant language functions of the right hemisphere? Prosody and emotional gesturing. *Archives of Neurology*, *36*, 144–148.

Ryalls, J. (1986). What constitutes a primary disturbance of speech prosody? A reply to Shapiro and Danly. *Brain and Language*, *29*, 183–187.

Ryalls, J., Joanette, Y., & Feldman, L. (1987). An acoustic comparison of normal and right-hemisphere-damaged speech prosody. Tapuscrit CHCN Working Papers 4.

Sabourin, L., Goulet, P., & Joanette, Y. (1988). La disponibilité lexicale chez les cérébrolésés droits. *Canadian Psychology\Psychologie Canadienne*, *29*, 2a.

Safer, M.A., & Leventhal, H. (1977). Ear differences in evaluation emotional tones of voice and verbal content. *Journal of Experimental Psychology: Human Perception and Performance*, *3*, 75–82.

Saffran, E.M., Bogyo, L.C., Schwartz, M.F., & Marin, O.S.M. (1980). Does deep dyslexia reflect right-hemisphere reading? In M. Coltheart, K. Patterson, & J.C. Marshall (Eds.), *Deep dyslexia* (pp. 381–406). London: Routledge and Kegan Paul.

Schiepers, C. (1980). Response latency and accuracy in visual word recognition. *Perception and Psychophysics*, *27*, 71–89.

Schlanger, B.B. (1973). Identification by normals and aphasic subjects of meaningful and meaningless emotional toned sentences. *Acta Symbolica*, *4*, 30–38.

Schlanger, B.B., Schlanger, P., & Gertsman, L.J. (1976). The perception of emotionally toned sentences by right hemisphere-damaged and aphasic subjects. *Brain and Language*, *3*, 396–403.

Schmit, V., & Davis, R. (1974). The role of hemispheric specialization in the analysis of Stroop stimuli. *Acta Psychologia*, *38*, 149–158.

Schmuller, J., & Goodman, R. (1979). Bilateral tachistoscopic perception, handedness and laterality. *Brain and Language*, *8*, 81–91.

Schuell, H. (1965). *Differential diagnosis of aphasia with the Minnesota test.* Minneapolis: University of Minnesota Press.

Schulhoff, C., & Goodglass, H. (1969). Dichotic listening side of injury and cerebral dominance. *Neuropsychologia*, *7*, 149–160.

Schweiger, A., Fiel, T., Dobkin, B., & Zaidel, E. (1983). Right hemisphere control of reading following left hemisphere damage: A case study of a patient with deep dyslexia and deep agraphia. Paper presented at B.A.B.B.L.E., Niagara Falls, Ontario.

Searle, J.R. (1969). *Speech acts.* Cambridge: Cambridge University Press.

Searle, J.R. (1982). *Sens et expression.* Paris: Editions de Minuit.

Searleman, A. (1977). A review of right hemisphere linguistic capabilities. *Psychological Bulletin, 84*, 503–528.

Searleman, A. (1983). Language capabilities of the right hemiphere. In A.W. Young (Ed.), *Functions of the right cerebral hemisphere* (pp. 87–111). London: Academic Press.

Seashore, C., Lewis, C., & Setveit, J. (1960). Seashore measures of musical talents. New York: Psychological Corp.

Segalowitz, S.J., & Bryden, M.P. (1983). Individual differences in hemispheric representation of language. In S.J. Segalowitz (Ed.), *Language functions and brain organization* (pp. 341–372). New York: Academic Press.

Semmes, J. (1968). Hemispheric specialization: A possible cue to mechanism. *Neuropsychologia, 6*, 11–26.

Sergent, J. (1983). Unified response to bilateral hemispheric stimulation by a split-brain patient. *Nature, 305*, 800–802.

Sergent, J., & Hellige, J.B. (1986). Role of input factors in visual-field asymmetries. *Brain and Cognition, 5*, 174–199.

Seron, X. (1979). *Aphasie et neuropsychologie.* Brussels: Mardaga.

Seron, X., Van Der Kaa, M.A., Van Der Linden, M., & Remits, A. (1982). Decoding paralinguistic signals: Effects of semantic and prosodic cues on aphasics' comprehension. *Journals of Communication Disorders, 15*, 223–231.

Seron, X., & Van Der Linden, M. (1979). Vers une neuropsychologie humaine des conduites émotionnelles. *L'Année Psychologique, 79*, 229–252.

Shallice, T., & Warrington, E.K. (1980). Single and multiple component central dyslexic syndromes. In M. Coltheart, K. Patterson, & J.C. Marshall (Eds.), *Deep dyslexia* (pp. 119–145). London: Routledge and Kegan Paul.

Shanon, B. (1979). Lateralization effects in lexical decision task. *Brain and Language, 8*, 380–387.

Shapiro, B.E., & Danly, M. (1985). The role of the right hemisphere in the control of speech prosody in propositional and affective contexts. *Brain and Language, 25*, 19–36.

Shewan, C., & Kertesz, A. (1984). Effects of speech and language treatment on recovery from aphasia. *Brain and Language, 23*, 272–299.

Sidtis, J.J., & Gazzaniga, M.S. (1983). Competence versus performance after callosal section: Looks can be deceiving. In J.B. Hellige (Ed.), *Cerebral hemisphere asymmetry. Method, theory and application* (pp. 152–176). New York: Praeger.

Sidtis, J.J., Volpe, B.T., Wilson, D.H., Rayport, M., & Gazzaniga, M.S. (1981). Variability in right hemisphere language function after callosal section: Evidence for a continuum of generative capacity. *Journal of Neuroscience, 1*, 323–331.

Smith, A. (1966). Speech and other functions after left hemispherectomy. *Journal of Neurology, Neurosurgery and Psychiatry, 29*, 467–471.

Sparks, R., Goodglass, H., & Nickel, B. (1970). Ipsilateral versus contralateral extinction in dichotic listening resulting from hemisphere lesions. *Cortex, 6*, 249–260.

Sperry, R.W. (1968). Hemisphere deconnection and unity in conscious awareness. *American Psychologist, 23*, 723–733.

Sperry, R.W. (1982). Some effects of disconnecting the cerebral hemispheres. *Science, 217*, 1223–1226.

Sperry, R.W., & Gazzaniga, M.S. (1967). Language following surgical disconnection of the hemispheres. In C.H. Millikan & F.L. Darley (Eds.), *Brain mechanisms underlying speech and language* (pp. 108–121). New York: Grune and Stratton.

Sperry, R.W., Zaidel, E., & Zaidel, D. (1979). Self recognition and social awareness in the disconnected minor hemisphere. *Neuropsychologia*, *17*, 153–166.

Stachowiak, F.J., Huber, W., Poeck, K., Kerschensteiner, M. (1977). Text comprehension in aphasia. *Brain and Language*, *4*, 177–195.

Stanovich, K.E., & Bauer, D.W. (1978). Experiments on the spelling-to-sound regularity effect in word recognition. *Memory and Cognition*, *6*, 410–415.

Stroop, J.R. (1935). Studies of interference in serial verbal reactions. *Journal of Experimental Psychology*, *18*, 643–662.

Sugishita, M. (1978). Mental association in the minor hemisphere of a commissurotomy patient. *Neuropsychologia*, *16*, 229–232.

Suls, J.M. (1983). Cognitive processes in humour appreciation. In P.E. McGhee & J.H. Goldstein (Eds.), *Handbook of humor research* (pp. 39–55). New York: Springer-Verlag.

Swisher, L.P., & Sarno, M.T. (1969). Token test scores of three matched patient groups: Left brain-damaged with aphasia, right brain-damaged without aphasia, non-brain-damaged. *Cortex*, *5*, 264–273.

Teuber, H.L. (1975). Recovery of function after brain injury in man. In R. Porter & D.W. Fitzsimmons (Eds.), *Outcome of severe damage to the central nervous system* (pp. 159–186). Amsterdam: Elsevier/Excerpta Medica.

Théroux, A.M. (1987). Trouble de la prosodie résultant d'une lésion hémisphérique droite: Une étude de cas. Unpublished Master Thesis, Université de Montréal, Montréal.

Thiery, E., Dietens, E., & Vandereecken, H. (1982). La récupération spontanée: Ampleurs et limites. In X. Seron & C. Laterre (Eds.), *Rééduquer le cerveau* (pp. 33–43). Brussels: Mardaga.

Toglia, M.P., & Batig, W.F. (1978). *Handbook of semantic word norms*. Hillsdale, N.J.: Lawrence Erlbaum Associates.

Tompkins, C., & Mateer, C.A. (1985). Right hemisphere appreciation of prosodic and linguistic indications of implicit attitude. *Brain and Language*, *24*, 185–203.

Tompkins, C.A., & Flowers, C.R. (1985). Perception of emotional intonation by brain-damaged adults: The influence of task processing levels. *Journal of Speech and Hearing Research*, *28*, 527–538.

Tompkins, C.A., & Jackson, S.T. (1988). Lexical connotation and right brain damage: Evidence from priming studies. Paper presented at the 26th Annual Meeting of the Academy of Aphasia, Montréal.

Tsao, Y.C., Feustel, T., & Soseos, C. (1979). Stroop interference in the left and right visual fields. *Brain and Language*, *8*, 367–371.

Tucker, D.M., Watson, R.T., & Heilman, K.M. (1977). Discrimination and evocation of affectively intoned speech in patients with right parietal disease. *Neurology*, *27*, 947–950.

Turner, S., Miller, L.K. (1975). Some boundary conditions for laterality effects in children. *Developmental Psychology*, *11*, 342–352.

Underwood, G. (1976). Semantic interference from unattended printed words. *British Journal of Psychology*, *67*, 327–338.

Underwood, G. (1977). Attention, awareness, and hemispheric differences in word recognition. *Neuropsychologia*, *15*, 61–67.

Underwood, G., & Whitfield, A. (1985). Right-hemisphere interactions in picture-word processing. *Brain and Cognition*, *4*, 273–286.

Underwood, P.J., Rusted, J., & Thwaites, S. (1983). Parafoveal words are effective in both hemifields: Preattentive processing of semantic and phonological codes. *Perception*, *12*, 213–221.

Urcuioli, P.J., Klein, R.M., & Day, J. (1981). Hemispheric differences in semantic processing: Category matching is not the same as category membership. *Perception and Psychophysics*, *29*, 343–351.

Vaid, J. (1983). Bilingualism and brain laterialization. In S.J. Segalowitz (Ed.), *Language functions and brain organization* (pp. 315–339). New York: Academic Press.

Van Dijk, T.A., & Kintsch, W. (1983). *Strategies of discourse comprhension*. New York: Academic Press.

Van Lanckner, D. (1980). Cerebral lateralizations of pitch cues in the linguistic signal. *Papers in Linguistics. International Journal of Human Communication*, *13*, 201–277.

Vargha-Khadem, F., O'Gorman, A.M., & Watters, G.V. (1985). Aphasia and handedness in relation to hemispheric side, age at injury and severity of cerebral lesion during childhood. *Brain*, *108*, 677–696.

Villardita, C., Grioli, S., & Quattropani, M.C. (1988). Concreteness/Abstractness of stimulus-words and semantic clustering to right brain-damaged patients. *Cortex*, *24*, 563–571.

Volpe, B.T., Sidtis, J.J., Holtzman, J.D., Wilson, D.H., & Gazzaniga, M.S. (1982). Cortical mechanisms involved in praxis: Observation following partial and complete section of the corpus callosum. *Neurology*, *32*, 645–650.

Von Mayendorff, N. (1911). *Die aphasischen Symptome*. Leipzig: Engelman.

Wada, J. (1949). A new method for the determination of the side of cerebral speech dominance. A preliminary report on the intracarotid injection of sodium amytal in man. *Medical Biology*, *14*, 221.

Wada, J., & Rasmussen, T. (1960). Intracarotid injection of sodium amytal for the lateralization of cerebral speech dominance. Experimental and clinical observations. *Journal of Neurosurgery*, *17*, 266–282.

Walker, E., & Ceci, S.J. (1985). Semantic priming effects for stimuli presented to the right and left visual fields. *Brain and Language*, *25*, 144–159.

Wapner, W., Hamby, S., & Gardner, H. (1981). The role of the right hemisphere in the apprehension of complex linguistic materials. *Brain and Language*, *14*, 15–33.

Warrington, E.K., & Shallice, T. (1980). Word form dyslexia. *Brain*, *103*, 99–112.

Wechsler, A.F. (1973). The effect of organic brain disease on recall of emotionally charged versus neutral narrative texts. *Neurology*, *23*, 130–135.

Weinstein, E.A. (1964). Affections of speech with lesions of the nondominant hemisphere. *Research Publications of the Association for Research in Nervous and Mental Disease*, *42*, 220–228.

Weinstein, E.A., & Kahn, R.L. (1955). *The denial of illness: Symbolic and physiological aspects*. Springfield, Ill.: Thomas.

Weinstein, E.A., & Keller, N.J.A. (1963). Linguistic patterns of misnaming in brain injury. *Neuropsychologia*, *1*, 79–90.

Weinstein, E.A., & Lyerly, O.G. (1972). Language in the visual hemi-inattention syndrome. *Transactions of the American Neurological Association*, *97*, 354–355.

Weinstein, E.A., & Puig-Antich, J. (1974). Jargon and its analogues. *Cortex*, *10*, 75–83.

Weintraub, S., Mesulam, M.M., & Kramer, L. (1981). Disturbances in prosody: A right hemisphere contribution to language. *Archives of Neurology*, *38*, 742–744.

Weniger, D. (1978). The recognition of stress patterns as a lexical and syntactic cue. Paper presented at the 16th Annual Meeting of the Academy of Aphasia, Chicago.

Weniger, D. (1984). Dysprosody as part of the aphasic language disorder. In F.C. Rose (Ed.), *Advances in neurology: Vol. 42. Progress in aphasiology* (pp. 41–50). New York: Raven.

Wepman, J.M., & Jones, L.V. (1961). *Studies in Aphasia: An Approach to Testing.* Chicago: University of Chicago.

Weylman, S.T., Brownell, H.H., & Gardner, H. (1986). Comprehension of indirect requests by organic patients. Paper presented at the Annual Meeting of the Academy of Aphasia, Nashville.

Whitaker, H.A., & Ojemann, G.A. (1977). Lateralization of higher cortical functioning: A critique. *Annals of the New York Academy of Sciences, 299*, 459–473.

Wilcox, M.J., Davis, G.A., & Leonard, L.B. (1978). Aphasic's comprehension of contextually conveyed meaning. *Brain and Language, 6*, 362–377.

Wilkins, A., & Moscovitch, M. (1978). Selective impairment of semantic memory after temporal lobectomy. *Neuropsychologia, 16*, 73–79.

Williams, C.E., & Stevens, K.N. (1972). Emotions and speech: Some acoustical correlates. *Journal of the Acoustical Society of America, 52*, 1238–1250.

Williams, P.C., & Parkin, A.J. (1980). On knowing the meaning of words which we are unable to report: Confirmation of a guessing explanation. *Quarterly Journal of Experimental Psychology, 32*, 101–107.

Winner, E., & Gardner, H. (1977). The comprehension of metaphor in brain-damaged patients. *Brain, 100*, 717–723.

Woods, B.T., & Carey, S. (1979). Language deficits after apparent clinical recovery from childhood aphasia. *Annals of Neurology, 6*, 405–409.

Woods, B.T., & Teuber, H.L. (1973). Early onset of complementary specialization of cerebral hemispheres in man. *Transactions of the American Neurological Association, 98*, 113–117.

Woods, B.T., & Teuber, H.L. (1978). Changing patterns of childhood aphasia. *Annals of Neurology, 3*, 273–280.

Wuillemin, D.B., Krane, R.V., & Richardson, B.L. (1982). Hemispheric differences in picture-word interference. *Brain and Language, 16*, 121–132.

Yeni-Komshian, G.H., & Rao, P. (1980). Speech perception in right and left CVA patients. Paper presented at the International Neuropsychological Society, San Francisco.

Young, A.W. (1982). Methodological and theoretical bases of visual hemifield studies. In J.G. Beaumont (Ed.), *Divided visual field studies of cerebral organization* (pp. 11–27). New York: Academic Press.

Young, A.W., & Ellis, A.W. (1985). Different methods of lexical access for words presented in the left and right visual hemifields. *Brain and Language, 24*, 326–358.

Young, A.W., Ellis, A.W., & Bion, P.J. (1984). Left hemisphere superiority for pronounceable nonwords, but not for unpronounceable letter strings. *Brain and Language, 22*, 14–25.

Zaidel, E. (1976). Auditory vocabulary of the right-hemisphere following brain bisection or hemidecortication. *Cortex, 12*, 191–211.

Zaidel, E. (1978a). Auditory language comprehension in the right hemisphere following cerebral commissurotomy and hemispherectomy: A comparison with child language and aphasia. In A. Caramazza & E.B. Zurif (Eds.), *Language acquisition and language breakdown* (pp. 229–275). Baltimore: Johns Hopkins University Press.

Zaidel, E. (1978b). Concepts of cerebral dominance in the split brain. In P.A. Buser & A. Rougel-Buser (Eds.), *Cerebral correlates of conscious experience* (pp. 263–284). Amsterdam: Elsevier–North-Holland Biomedical Press.

Zaidel, E. (1978c). Lexical organization in the right hemisphere. In: P.A. Buser & A. Rougel-Buser (Eds.), *Cerebral correlates of conscious experience* (pp. 177–197). Amsterdam: Elsevier–North-Holland Biomedical Press.

Zaidel, E. (1982). Reading by the disconnected right hemisphere: An aphasiological perspective. In Y. Zotterman (Ed.), *Dyslexia: Neuronal, cognitive and linguistic aspects* (pp. 67–91). Oxford: Pergamon Press.

Zaidel, E. (1983a). Disconnection syndrome as a model for laterality effects in the normal brain. In J.B. Hellige (Ed.), *Cerebral hemisphere asymmetry. Method, theory, and application* (pp. 95–151). New York: Praeger.

Zaidel, E. (1983b). A response to Gazzaniga. Language in the right hemisphere convergent perspectives. *American Psychologist, 38*, 542–546.

Zaidel, E. (1985). Language in the right hemisphere. In D.F. Benson & E. Zaidel (Eds.), *The dual brain* (pp. 205–231). New York: Guilford Press.

Zaidel, E. (1986a). Callosal dynamics and right hemisphere language. In F. Lepore, M. Ptito, & H.H. Jasper (Eds.), *Two hemispheres: One brain* (pp. 435–462). New York: Alan R. Liss.

Zaidel, E. (1986b). Hemispheric locus of lexical congruity effects: A neuropsychological reinterpretation of psychololinguistics results. Paper presented at the 9th European Conference of the International Neuropsychological Society, Veldhoven.

Zaidel, E., & Peters, A.M. (1981). Phonological encoding and ideographic reading by the disconnected right hemisphere: Two case studies. *Brain and Language, 14*, 204–234.

Zaidel, E., & Schweiger, A. (1983). Right hemisphere contribution to reading: The case of deep dyslexia. Paper presented at the annual meeting of the Academy of Aphasia, Minneapolis.

Zaidel, E., & Schweiger, A. (1984). One wrong hypothesis about the right hemisphere: Commentary on K. Patterson and D. Besner, "Is the right hemisphere literate?" *Cognitive Neuropsychology, 1*, 351–364.

Zangwill, O.L. (1967). Speech and the minor hemisphere. *Acta Neurologica Belgica, 67*, 1012–1020.

Zollinger, R. (1935). Removal of the left cerebral hemmisphere. *Archives of Neurology and Psychiatry, 34*, 1055–1064.

Zurif, E. (1974). Auditory lateralization: Prosodic and syntactic factors. *Brain and Language, 1*, 391–404.

Zurif, E., & Mendelsohn, M. (1972). Hemispheric specialization for the perception of speech sounds: The influence of intonation and structure. *Perception and Psychophysics, 11*, 329–332.

Zurif, E., & Sait, P.E. (1970). The role of syntax in dichotic listening. *Neuropsychologia, 8*, 239–244.

Author Index[1]

[1]Many thanks to Marianne Corre and to Lucien for their help in the preparation of this index.

Subject Index

Many thanks to Colette Cerny as well as to Madeleine Samson for their help in the preparation of this index.